Organizational RESOURCE MANAGEMENT

THEORIES, METHODOLOGIES, & APPLICATIONS

Organizational RESOURCE MANAGEMENT

THEORIES, METHODOLOGIES, & APPLICATIONS

Jussi Kantola

CRC Press
Taylor & Francis Group
Boca Raton London New York

CRC Press is an imprint of the
Taylor & Francis Group, an **informa** business

CRC Press
Taylor & Francis Group
6000 Broken Sound Parkway NW, Suite 300
Boca Raton, FL 33487-2742

Printed on acid-free paper
Version Date: 20150909

International Standard Book Number-13: 978-1-4398-5120-3 (Hardback)

Visit the Taylor & Francis Web site at
http://www.taylorandfrancis.com

and the CRC Press Web site at
http://www.crcpress.com

This book is dedicated to all those who can see
the world without sharp boundaries.

Contents

List of Figures

List of Tables

Abstract

The management of organizational resources is difficult! Managers face serious challenges when they are managing the required resources as properly as they can for the benefit of their organization. Organizational resources are management objects that represent themselves with challenges, such as abstract and complex nature, lack of holistic view, multiobjectiveness, low degree of motivation and participation, etc. This book presents an approach and a system that aims to tackle these management challenges. The approach is based on four propositions that together form a solid framework for the management of organizational resources. The system supports the approach in computational practice.

The propositions state that organizational resources that are considered as management objects must first be conceptually modeled. With the help of these models, stakeholders are integrated into the management process. The meaning of how different individuals and stakeholders perceive the organizational resources they are dealing with is computed using soft-computing technology. The computed meaning is used as an important knowledge input into management decision-making process together with strategy and other relevant material. Targeted and justified actions regarding organizational resources are planned and executed in periodic management cycles. It is possible to manage organizational resources in a completely new uniform way. This proposed management approach captures and utilizes situational perceptions and aspirations of individuals and stakeholders for the better future of the organizational resources and the whole organization.

This approach has been used and tested in tens of companies, organizations, and universities in more than ten languages around the world. Based on the broad experience of the approach, it is safe to say that the theory-based and practically tested evolute approach and system presented in this book for the management of organizational resources can be safely utilized widely in the field of management study and practice. The goal is to develop relevant organizational resources and, thus, improve the performance of the whole organization in an efficient and targeted manner.

Keywords: Evolute, coevolute, organizational resource, concept, management object, metaknowledge, system, ontology, instance, perception, fuzzy logic, soft computing

Preface

This work was initiated at the HAAMAHA (Human Aspects of Advanced Manufacturing Agility and Hybrid Automation) conference in Krakow, Poland, in 2000, where I met Professor Hannu Vanharanta from the Tampere University of Technology, Pori unit, Finland. At that time, I was working for ABB in Finland. During the conference, we talked about our research and the work that we had conducted earlier. The idea of cooperating with each other to combine our previous works was born while we were eating delicious steaks. The idea was to apply my fuzzy logic–related doctoral research (Kantola 1998) to Professor Vanharanta's research in the area of strategic management (Vanharanta 1995). Previously, I had applied fuzzy logic to the evaluation of computer-integrated manufacturing, organization, and people system design (The CIMOP application) together with Professor Waldemar Karwowski at the University of Louisville, Kentucky. The purpose of that work was to evaluate the existing CIM installations and investment and other plans for future CIM installations in order to find specific areas in CIMOP system designs that needed improvement. Professor Vanharanta's earlier research dealt with hyperknowledge and continuous strategy in executive support systems based on fundamental components of any organization: capital, work, and people.

The basic approach in both of our research was to look at organizational objects in a holistic manner. Also, we both attempted to primarily help management at their work. These aspects, together with the exciting possibilities enabled by fuzzy logic, motivated us to start preparing for our first cooperative project proposal for TEKES (Finnish Innovation Fund). The proposal was approved, and we were able to start the work. The first fuzzy logic computing application we created was named FuzzyLeader; it was based on the same fuzzy computing principles as the CIMOP system. Since that time (2002), the number of projects and people working on this initial idea and its methods and applications has increased significantly in Finland and internationally. Tens of MSc and many PhD theses have been done, and are being done, lots of articles and book chapters have been published, and many workshops, sessions, presentations, and even conferences have been conducted in many countries on this topic. Thousands of people in different kinds of academic, municipal, and industrial contexts have already used and applied the approach using many different languages, thus also enabling us to look at the management of the meaning of organizational and business objects according to cultural differences in a new, justified, and focused way. This book is another formal result of this cooperation.

The first version of this work was prepared as my doctoral thesis at the Tampere University of Technology in Finland in 2005 (Kantola 2005). The idea of writing the content of the thesis as a book was in the air for few years, and finally in early 2010, I proposed it to the publisher—CRC Press—and they accepted. Many years of experience and feedback from this research and its applications in different kinds of organizations gave me an opportunity to improve and update the text of the original thesis and write the content in a less technical and academic way—at least that was the hope.

We humans are quite used to developing metrics and measurement tools in technical domains, but as soon as we start dealing with humans and human systems (systems that involve humans), reality gets rather complicated, and we have to ask ourselves, "Can we really reduce human systems to precise numbers?" The answer in this book is, No! Management must deal with human systems on a daily basis in companies and organizations. In reality, we simply cannot leave nonprecise humans out of everyday management work and hope for the best, or leave the human resource management function to take care of "those human operations."

This book attempts to put forward an approach on how to model complex and abstract organizational resources, such as humans and human systems, and how to develop and manage these resources for the benefit of the organization and its employees by integrating employees and other stakeholders into the practical management work. For that purpose, relevant concepts in a work domain must first be clarified so that they can be understood and used by people who are doing their work in that domain. Although we cannot directly measure abstract resources, we certainly can perceive various aspects of the resources in our everyday work. Every stakeholder views the concepts that describe the resources differently, and every individual has their own experience and perceptions of the resources in real life. The overall context combined with the context of the perceiver surely result in different kinds of perceptions of the resources. The collective and conscious experience in normal working situations is very valuable for an organization as it can provide an important bottom-up understanding of the organizational resources in different situations and provide guidance as to what should be done to deal with the resources in a timely, dynamic, and continuous manner. To be able to know and use that hands-on experience will greatly help management. But, how does one capture people's perception of abstract concepts and turn that valuable knowledge input into management and development actions and real resources in a timely, dynamic, and continuous manner?

The new management approach presented in this book is called the evolute approach. The evolute approach preciciates the meaning of organizational resources for management on three levels: concept level, organizational resource level, and situation level. The term "preciciation" is adopted from Professor Zadeh, the father of fuzzy logic, who coined the term in the contexts of CWW (computing with words) and GTU (generalized theory of uncertainty) (Zadeh 2005). In these contexts, it refers to preciciating the meaning of words, or making the meaning of words more precise, so that the meaning can be "connected" to machines and computing systems. Only after the preciciation of the meaning can we address problems described in natural language using computing. Similarly, in the field of management, the reality is that we have to address problems and situations that are perceived and described in natural language at workplaces. The evolute approach helps managers to lead their people and manage their resources in a more preciciated way. If leadership and management efforts were justifiably better channeled, we would see better results in organizations with less effort.

The scope of the book is general management since the evolute approach is applicable to any kind of management field. It can be applied to the management of myriad objects—not only organizational resources. This book explains the principles of the evolute approach in theory and practice. Other existing approaches that are used to manage organizational resources are also viewed and compared to the management framework presented in this book.

This book consists of 10 chapters organized in the following order: Chapters 1 and 2: organizational resources management framework; Chapter 3: the evolute approach and system to manage organizational resources; Chapters 4 through 7: four real-life cases where evolute tools were used; Chapter 8: other existing approaches and methodologies that are used to manage organizational resources; Chapter 9: resource innovation based on the evolute approach; and Chapter 10: discussion and future work.

I hope that you find something interesting in this book that will be helpful to you and your work! You are also welcome to join us in this new field of management work.

Jussi I. Kantola
Vaasa, Finland

REFERENCES

Kantola, J. 1998. *A Fuzzy Logic Based Tool for the Evaluation of Computer Integrated Manufacturing, Organization and People System Design*. Louisville, KY: University of Louisville.

Kantola, J. 2005. Ingenious management. Doctoral thesis, Pori, Finland: Tampere University of Technology.

Vanharanta, H. 1995. *Hyper Knowledge and Continuous Strategy in Executive Support Systems*. Turku, Finland: Åbo Akademi.

Zadeh, L. 2005. Toward a generalized theory of uncertainty (GTU)––An outline. *Information Sciences* 172(1–2) (June 9).

Acknowledgments

First, I acknowledge those people who were involved in my PhD thesis work during 2002–2006. This book is based on that thesis. The research for the thesis was carried out at the Tampere University of Technology (TUT), Pori, Finland, and in its tiny satellite office in Turku during 2002–2005. I am grateful to Professor Hannu Vanharanta who was my supervisor. The thesis would not have been possible without his expertise, ideas, enthusiasm, and exceptional 24/7 support. The time I spent working with him on the thesis research was an educating and fun experience for me. I am also very grateful to Professor Waldemar Karwowski at the University of Central Florida (UCF), Orlando, who helped me a lot by providing several opportunities related to the research along the way. He also helped me think logically when I was writing my thesis in Louisville, Kentucky, during the summer of 2005. Professor Karwowski was my supervisor at the University of Louisville when I was developing the CIMOP system during 1996–1998. I thank Professor Yoon Chang who was the opponent at the defense of my dissertation in 2006 in Finland. I acknowledge the coauthors of the articles in the thesis for their very important contribution. Without them, the thesis in the format of article collection would not have been possible. Financially, TEKES (Finnish Innovation Fund) made the thesis possible by providing funding for the PMPM project (Dnro 1604/31/01), 4M project (Dnro 665/31/03), and Spede project (Dnro 267/31/04) at TUT, Pori, during 2002–2005. I am very grateful to TEKES for this. Suomen Kulttuurirahasto, Satakunnan Säätiö, provided financial support to me to finish my thesis. Thank you.

Second, I acknowledge all the people, companies, and organizations that have been involved in developing the evolute approach since 2006. I want to thank the energetic professor Hannu Vanharanta at TUT, who has a very broad understanding of what is needed and what should be done in the field of management. He has initiated and led most of the domain research since 2005 in the context of the approach that is described in this book. He is expanding this work constantly to new, important, and unexplored domains. I acknowledge the very important guidance provided by Professor Waldemar Karwowski at UCF. He has provided advice as to which direction the presented approach should be developed. He also continuously provides many opportunities to do this work globally—the latest being the first Co-Evolute Conference on Human Factors, Business Management and Society at Las Vegas in July 2015 at AHFE2015 (Advanced Human Factor and Ergonomics) Conference.

I thank all the researchers who have adopted the approach presented in this book for their research and have conducted the work in many universities and companies worldwide using different languages. I also thank researchers who have written conference articles, journal articles, and book chapters on this approach. All of them have provided extremely important feedback and critical and creative ideas on how to apply the approach and how to develop it further. I thank Evolute Research Centers (ERCs) around the world, which are implementing this research and work into practice in companies and universities. I thank the people who translated the domain content to other languages. This is a truly coevolutionary work.

I thank the students who have taken part in the research projects in Pori, Tampere, Turku, and Vaasa in Finland; Gerona and Barcelona in Spain; Poznan in Poland; and Daejeon and Seoul in Korea, to name few.

I thank Hannu Vanharanta, Jarno Einolander, Tero Reunanen, and Kari Ingman for writing the case studies included in this book.

I am grateful to my wife Sibel for her support during the very long period of making this book. Thank you, Sibel! I could not have done it without you. I also thank our sons Selim (13), Erhan (8), and Ville (1) for bringing happiness and action into our lives. They participated in this project in their own and different ways. Together, they present the toughest management challenge for us. My parents Leena and Markku and my sister Sanna encouraged me for years to finish this book. My parents also helped us with taking care of the kids at critical times. Thank you for that!

I am extremely grateful to CRC Press for accepting this book for publication. I appreciate the great help and patience from Cindy Carelli, Kari Budyk, and others who were patient with me and guided me throughout this project. Without them, this project would have been impossible. I am also grateful to all those who have contributed toward the completion of this book but not have been mentioned here. Thank you!

Author

Jussi I. Kantola is a professor in the Department of Production at the University of Vaasa, Finland. Before that, during 2009–2012, he was associate professor in the Knowledge Service Engineering Department at the Korea Advanced Institute of Science and Technology (KAIST). During 2003–2008, he worked in the Tampere University of Technology, Pori, Finland, and the University of Turku, Finland, in various research roles, including that of research director in the IE and IT departments. He earned his first PhD in industrial engineering from the University of Louisville, Kentucky, in 1998, and his second PhD in industrial management and engineering from Tampere University of Technology, Pori, Finland, in 2006. During 1999–2002, he worked as an IT and business process consultant in the United States and in Finland for Romac International and ABB. His current work interests include the evolute approach and system, design theories, and fuzzy and ontology applications.

Definitions Used in This Book

Circles of mind: A metaphor that was created from the Baars theater model (1997) as a circular theater model (Vanharanta 2003). The components and contents as such are the same, but the physical model, the metaphor, is different. Conscious experience is in the middle of the model, and the players and audience are situated around it. The spotlight controller, director, and the context form the outer circles that surround the audience systems.

Concept: The constituents of thoughts (Margolis and Laurence 2014). Something conceived in the mind (thought, notion), or an abstract or generic idea generalized from particular instances (*Merriam-Webster* 2014).

Creative tension: The difference between current reality and vision, objectively thinking (Senge 1990). Creative tension is considered for objects that are inside oneself.

Evolute approach: A management paradigm that systematically utilizes evolute system and cyclic process to manage organizational resources dynamically in companies and organizations.

Evolute system: An online soft-computing-based technology that supports the management of organizational resources according to the evolute approach.

Extrospection: Examination or observation of what is outside oneself—opposed to introspection (*Merriam-Webster* 2014).

Fuzzy logic: A system of reasoning and computation in which the objects of reasoning and computation are classes with unsharp (fuzzy) boundaries (Zadeh 2014).

Fuzzy set: A mathematical set with unsharp boundaries that maps the linguistic value of a variable to its numerical value (Zadeh 1965).

Indicator: Attributes that describe the concept in the object ontology. Perceivable evidence of the concepts is indicators. Indicators are linguistic variables that are expressed in natural language.

Ingenious management: Using object ontologies and instances to justify decisions and actions in an organization's management and development. Ingenious management allows actions to be efficiently directed to the right areas. This allows for a more efficient use of organizational resources and also enables the fast positioning and development of organizational resources (Kantola 2005).

Instance: The captured perception of the concepts in object ontology. Instance contains the views of current reality and future vision in specific context (Kantola 2005). Instance is a dataset with several attributes.

Instance matrix: A collection of Instance vectors (Kantola 2005).

Instance matrix—dynamic: Instance matrix over periods of time (Kantola 2005). Dynamic instance matrix can also be viewed as a 3D instance space, where each node within the space is an Instance as defined in this book.

Instance space—3D: Same as dynamic instance matrix.

Instance vector: A collection of several Instances (Kantola 2005).

Interpreter: A module in the evolute system that computes ontology-based meanings of perceptions.

Introspection: Examination or observation of what is inside oneself—opposed to extrospection (*Merriam-Webster* 2014). Learning about one's own currently ongoing, or perhaps very recently past, mental states or processes (Schwitzgebel 2014).

Knowledge increment: A new piece of knowledge that can be added to a knowledgebase on many different knowledge levels starting from the fuzzy sets of indicators to dynamic instance matrices.

Linguistic variable: Variables with linguistic values. Linguistic variables contain fuzzy sets to cover the whole universe of discourse of the variable. Indicators are linguistic variables.

Management objects (MO): Organizational resources are tangible or intangible management objects: physical objects, financial objects, mental objects, logical objects, abstract objects, etc., that are managed and developed by the management of an organization.

Management object ontology (MOO): An explicit specification of a conceptualization of the management object.

Ontology: An explicit specification of a conceptualization (Gruber 1993).

Ontology-based self-evaluation: The evaluation of the indicators of internal and external concepts in (organizational resource) ontologies.

Organizational resources (ORs): Tangible and intangible assets that an organization has and/or utilizes to achieve the goal: people, organization, production, operations, work, investments, cultures, etc.

Organizational resource innovation: Knowledge increments increase a chance to create and put in practice innovative new ways to develop and utilize organizational resources in specific contexts.

Preciciation: Making the meaning of words more precise so that the meaning can be "connected" to machines and computing systems (Zadeh 2005).

Proactive vision: The difference between current reality and vision, objectively thinking (Vanharanta et al. 2012; Senge 1990). Proactive vision is considered for objects that are outside oneself.

Proposition: The sharable objects of the attitudes and the primary bearers of truth and falsity (McGrath 2014). Research propositions are statements about the concepts presented in research that may be judged as true or false (Cooper and Schindler 2002).

Ontology repository: A database module in the evolute system that contains organizational resource ontologies and other domain ontologies as well. The repository also stores Instances that are used for management purposes.

System: A group of interacting, interrelated, or interdependent components that form a complex and unified whole (Anderson and Johnson 1997).

REFERENCES

Anderson, V. and Johnson, L. 1997. *Systems Thinking Basics—From Concepts to Causal Loops*. Waltham, MA: Pegasus Communications, Inc.

Baars, B.J. 1997. *In the Theater of Consciousness: The Workspace of the Mind*. Oxford: Oxford University Press.

Cooper, D.R. and Schindler, P.S. 2002. *Business Research Methods*. Boston, MA: Irwin McGraw Hill.

Gruber, T.R. 1993. A translation approach to portable ontologies. *Knowledge Acquisition* 5(2) (April): 199–220.

Kantola, J. 2005. Ingenious management. Doctoral thesis, Pori, Finland: Tampere University of Technology.

Margolis, E. and Laurence, S. 2014. Concepts. In Zalta, E.N. (ed.), *The Stanford Encyclopedia of Philosophy*, Spring 2014 edition. Stanford, CA: Stanford University.

McGrath, M. 2014. Propositions. In Zalta, E.N. (ed.), *The Stanford Encyclopedia of Philosophy*, Spring 2014 edition. Stanford, CA: Stanford University.

Merriam-Webster, 2014. An Encyclopedia Britannica Company. Accessed 3 January. http://www.merriam-webster.com/.

Schwitzgebel, E. 2014. Introspection. In Zalta, E.N. (ed.), *The Stanford Encyclopedia of Philosophy*, Spring 2014 edition.

Senge, P. 1990. *The Fifth Discipline: The Art & Practice of Learning Organization*. New York: Currency Doubleday.

Vanharanta, H., 2003. Circles of mind. In *Proceedings of Identity and Diversity in Organizations—Building Bridges in Europe. XIth European Congress on Work and Organizational Psychology*. Lisbon—EAWOP, Portugal.

Vanharanta, H., Magnusson, C., Ingman, K., Holmbom, A., and Kantola, J. 2012. Strategic knowledge services. In Kantola, J. and Karwowski, W. (eds.), *Knowledge Service Engineering Handbook*, pp. 527–555. Boca Raton, FL: CRC Press.

Zadeh, L. 1965. Fuzzy sets. *Information and Control* 8(3): 338–353.

Zadeh, L. 2005. Toward a generalized theory of uncertainty (GTU)—An outline. *Information Sciences* 172(1–2) (June 9).

Zadeh, L. 2014. [bisc-group] Soft boundaries. A message to BISC (Berkeley Initiative in Soft-Computing) email group, Berkeley, CA.

REFERENCES

Anderson, V. and Johnson, L. 1997. Systems Thinking Basics: From Concepts to Causal Loops. Waltham, MA: Pegasus Communications, Inc.

[illegible]

Cooper, D.R. and Schindler, P.S. [illegible]. Business Research Methods. [illegible] McGraw-Hill.

[illegible]

[illegible]

[illegible]

[illegible]

1 Introduction

This first chapter of the book examines the field of management from a specific angle. The idea is to explore the tasks that management is facing everyday from the point of view of its different challenges. Management activity does not take place in a laboratory or in some stable environment where everything is controlled and outcomes can easily be predicted. The reality is unknown and the future is unpredictable. The reality and the future at work do not follow the style of textbooks where specified problems relating to simplified objects in simplified contexts are solvable and certain answers are the correct ones. Such clear questions with clear answers do not exist in real life. The reality is different and much more complex than we are aware of. The purpose of this first chapter is to explore the management reality from different viewpoints that are very real. We all have some personal experiences of these viewpoints as well. These challenges are very real that organizations everywhere in the world face when trying to manage their resources as well as they can often without proper working solutions.

After we have looked into these real management challenges, we can search for different approaches to tackling these challenges. But first we have to explore what has to be done, and after that we will present different ways to do it. As a comparison, the design presented by Suh (1990, 2001) includes the "what" and "how" aspects of optional solutions, that is, first deciding what needs to be done, and then finding optional approaches to how it can be done. In this book, the goal is to design a working management solution for the given management challenge. This is done with the help of propositions that together form the big picture of the management field to meet these real challenges and to manage organizational resources in real life in real contexts. These propositions and the framework they form together represent the author's experience and understanding of the field from both the practitioner's and the researcher's viewpoints. But first, let us begin by looking at the challenges that are facing management.

There is management everywhere. All kinds of companies and organizations are being managed by a variety of different kinds of approaches and methods. Existing and new methods are used and tested constantly in organizations. In this book, organization refers to all kinds of organizations including companies, municipalities, nonprofit organizations, teams, groups, etc. Therefore, the scope of management challenges is huge—encompassing every organization on this planet.

It is quite easy to think that right decisions made in an organization will obviously have a positive impact on the organization, its people, its other resources, operations, and business. On the other hand, wrong, biased, delayed, or inadequate decisions will of course have a negative impact on the organization, its people, its other resources, operations, and business. Clearly, the quality of each decision makes a difference for the organization. If the degree of right decisions in the organization can be increased, it will likely have a positive effect on the organization. This surely

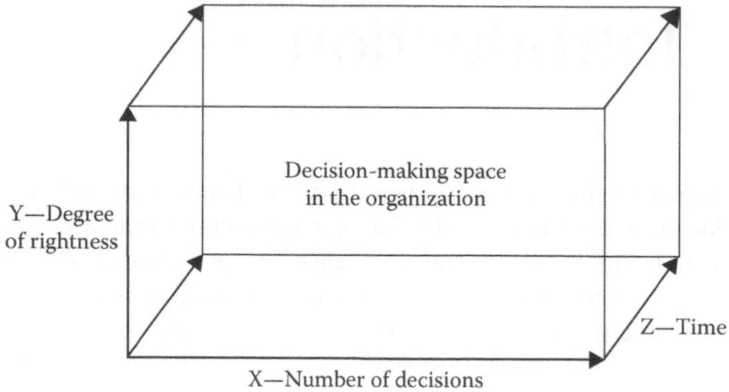

FIGURE 1.1 Simplified view of the decision-making space in the organization.

should be the goal of every single manager. This seems pretty simple and straightforward, but how can this be done in the organization? How can the degree of right decisions made in the organization be increased? It seems logical to think that the key lies in improving the degree of rightness of each single decision, and then integrating the new improved approach over all the decision-making space (Figure 1.1). Clearly, increasing the number of decisions without improving the degree of rightness of each decision would not lead to a good result. In Figure 1.1, Decision-making space = Number of decision × degree of rightness per decision × time. The X-axis shows the number of decisions per time period (e.g., per year); the Y-axis shows the degree of rightness of each decision; and the Z-axis brings a longitudinal view to the picture by looking at X and Y over a number of time periods. This 3D space can be empty (low degree) or full (high degree) or anything in between, depending on how the organization takes care of its decision making regarding its resources.

Certainly, the credibility of the knowledge input into the decision-making process can increase the degree of rightness of each decision, and therefore more decision-making space can be "filled in" the organization. Credibility in this context refers to more reliable understanding of the current reality and more reliable predictions of the future reality. But how to increase the credibility of knowledge input into the decision-making process in a real working and business environment to help the management make more right decisions? How to utilize explicit and tacit knowledge inputs (Nonaka and Takeuchi 1995; Nonaka et al. 2000)? How to utilize data streams, information streams, and human knowledge streams to generate new kinds of knowledge services that can help managers and other employees in their work (Kantola and Karwowski 2012; Vanharanta et al. 2012)?

At this point, we can make a pretty reliable statement that there are no simple solutions for complex management challenges. We can easily assume that there is more than one reason behind wrong, delayed, biased, and inadequate decisions in the organization. In the following paragraphs, some of these reasons are explored from different viewpoints, starting from the whole organization and then moving toward resources that the organization has and utilizes for the purpose of creating value for the customers.

From an organizational learning point of view, it is crucial that the organization learns at least at the same speed as its competitors. If the organization can develop itself faster than its competitors, its relative position in the dynamic business environment compared to its competitors will improve (Argyris 1999). The organization also has to be able to adapt to changes in its external and internal environment, whether they are Political, Economic, Social, or Technological (PEST), or any other kind of changes in the environment. There are models and templates available to perform PEST analyses of the environment of an organization. Other variations include PESTLE and PESTEL, where E stands for the Environment and L for Legal. This kind of contextual consideration can be seen as something similar to the "survival of the fittest" concept in evolutionary theory (Darwin 1869), where the survival of the fittest refers to the ability of individuals and species to adapt to changes in the environment. The winners are those who can adapt quickly to changes in the dynamic living environment. Such adaptability and faster development of the organization in an ever-changing dynamic operating and business environment requires right and timely decisions regarding the organization and its different kinds of resources. In this kind of activity, timely management decision making is the core element. Management decision making determines the future and fate of each organization. Many different approaches to management decision making have been developed.

Decision making in an organization concerns those assets, resources, and all other objects that the organization has and utilizes for its operations and business. If the organization can develop its resources faster to fit the requirements set by the environment than its competitors can develop their resources, the relative position of the organization in competition compared to its competitors' positions will improve in time. In other words, if the organization makes more right decisions regarding its resources than its competitors do regarding their resources, the relative position of the organization in competition compared to its competitors' positions will improve in time. Thus, *the goal of organizational resource management is right and timely decision making to develop relevant organizational resources in their real contexts.* We can retain this commonsense reasoning as the motivation for this book. It provides the reason for the content and structure of the book. From the management point of view, this goal means that the *Right Decision Principles* shown in Table 1.1 should be followed when a decision is made. This

TABLE 1.1
Right Decision Principles Provide the Guideline for the Development of Organizational Management Approaches

Right Decision Principles
1. The most reliable explicit knowledge of the object[a] is available
2. The most reliable tacit knowledge of the object is available
3. The most reliable understanding of the context and situation of the object is available

[a] Object in this book refers to organizational resource.

also means that when different methodologies, approaches, and tools are developed for organizational resource management, these principles should form a leading guideline.

Management decides where and how to direct efforts to develop the resources of the organization as time passes. From this perspective, this book aims to find practical solutions to help management to manage organizational resources for the benefit of the organization, its employees, and customers. The so-called typical speed of development of organizational resources may be accelerated by improving the degree of right decisions in a proactive manner. Such right decisions may be focused and targeted decisions at a certain point of time. In the next section, the nature of organizational resources that generate value for the organization's customers will be examined. Perhaps, the nature of resources reveals something to us about how we should approach them from the management action perspective so that the right decision principles listed in Table 1.1 are met when important decisions are made.

1.1 NATURE OF ORGANIZATIONAL RESOURCES

Let us start by considering what the resources or assets are that organizations may have and what the resources or assets are that the organization can utilize in its operations and business. In the upper image in Figure 1.2, we can see an office space where people are doing their work. This can be an engineering office where products and services are designed and developed. It may be a sales office where products and services are sold. Or it may be some other kind of office. In the second image, we can see a plant that could be a power plant, a chemical plant, or some other kind of plant. These organizations utilize different kinds of resources to produce and make products and services for customers as well as to add value to customers and to customers' businesses. What resources are needed in each case depends on the organization, its business area, its products and services, as well as the time and situation at hand. This means that different sets of resources are needed at different times. Therefore, the set of necessary resources in the organization is dynamic. In the images, we can see people, that is, human or intellectual resources, machinery, computers, information systems, investments, facilities, work processes, production, operations, and organizations. An organization needs these different kinds of resources to create value. All such resources are called organizational resources in this book.

Common to these two images is the fact that the most of these organizational resources may seem obvious concepts. But that is not all. Investments portfolios, different kinds of processes such as learning processes, RDI (research, development, and innovation) and sales processes, logistic solutions, patents, cultures, subcultures, time, etc., also belong to organizational resources. One typical way to look at different types of resources is to classify them into tangible and intangible resources (Berry 2005); in other words, what we can see (tangible) and what we cannot see (intangible). There is an old saying, "out of sight, out of mind," that is actually quite troubling for management decision making. The invisible resources can be the most difficult ones to manage.

All organizational resources are quite complex systems involving several elements that need to interact with each other. These interacting elements may be technical

(a)

(b)

FIGURE 1.2 (a) An office and (b) a plant at work.

elements, human elements, or some abstract and vague conceptual elements, that is, invisible elements. Systems, subsystems, and the derivatives of other organizational resources are also considered as organizational resources in this book. Any concepts that are based on or derived from tangible and intangible organizational resources are also considered as organizational resources in this book.

When we think of human elements and their interaction, we enter into a more difficult and invisible side of organizational resources. If we needed to show what organizational resources are and what they actually mean, we could not do it comprehensively. Are these obvious, visible resources, such as those seen in Figure 1.2, so obvious after all? Another question is, are there some other organizational resources that are as yet unrecognized and thus invisible to us? It is likely there are invisible organizational resource-related elements "present" in organizations that can make the difference between success and a failure. For example, it is likely there are intra- and interorganizational systems that are part of value creation processes that are not yet revealed by research. As the world changes, new kinds of organizational resources will emerge (or will be revealed) and some may become obsolete for some organizations.

In this book, organizational resources are considered as the objects of management decision making. They are affected by decisions made in the organization. Many of these organizational resources are present in companies every day and "touch" managers' daily work. Organizational resources are a great challenge for managers and for decision making in general. Are managers and personnel aware of their organizational resources today? Using top-down thinking, the first organizational resource question that must be answered by the organization, therefore, is explained in Section 1.1.1.

1.1.1 What Are Our Organizational Resources Today?

It is very important that an organization can answer this question. Otherwise, it is clearly not possible to manage the organization. After answering the first question, we can start looking into the resources and try to get clues for how to approach them from a management perspective. Organizational resources create very real difficulties for organizations and their management due to their characteristics. When we examine "the nature of organizational resources," we can hopefully find clues as to which direction to go to improve the right decision rate.

It is hard to imagine organizational resources without people. There have been attempts to fully automate factories, but so far these attempts have not been successful. It is likely these efforts will not be successful in the near future. Let us accept the fact that humans are an integral part of organizational resources. Humans as part of organizational resource systems mean that humans are on the input side and also on the output side of these systems. Humans are both causes and effects in organizational resource systems. If we try to ignore humans as part of organizational resource causality, we cannot be successful. There are so many models in management, business, and engineering where humans are completely missing. Such simplistic approaches will not succeed in the real-life dynamic operational and business environment.

The more complex and abstract resources there are in the organization, the less existing working solutions we can find that we can actually utilize in practice. If the existence of a resource is already recognized in general, we may find theories, but no practical tools may be available. If we consider a newly recognized organizational resource system, we may not even find existing theories. We can quickly notice that all these complex and abstract "new" resources typically include human elements. Humans clearly are integral parts of those organizational resources that do not yet have good working solutions and are the most difficult cases for management.

In the technical elements of organizational resources, we can reach the required precision levels in operations by measuring variables and adjusting parameters. For example, we can apply statistical methods to monitor statuses of the elements or of the whole resource and take some action if a certain control limit is exceeded. Management is a kind of measurement and adjustment activity in relation to organizational resources as part of operational processes. We can do this kind of activity with the help of computers and we can also automate our processes. This is typical engineering work.

TABLE 1.2
Nature of the Technical and Human Elements of Organizational Resources

Technical Elements	Human Elements
Precision	Imprecision
Measurement	Evaluation
Certainty	Uncertainty
Variables (can be measured)	Concepts (can be perceived)
Carry value	Carry meaning
Predictability	Unpredictability
Easier to manage	More difficult to manage
Less potential unleashed in management field	More potential unleashed in management field

On the other hand, we cannot measure and effectively manage imprecision and uncertainty associated with people. From the management perspective, the technical elements of organizational resources are more predictable and manageable than human elements. It is easier to manage and adjust technical elements. At the same time, however, we know that if we want to be successful we cannot ignore all imprecise and uncertain elements of organizational resources. We have to work also with imprecision and uncertainty related to organizational resources. In place of measurable variables in technical elements, we have to take concepts that are only present in human minds, such as individual and shared perceptions and views of objects. From the management perspective, human elements are much more difficult to manage than technical elements. The greatest potential, to the author's mind, and difficulty in the field of management is related to these human elements. Table 1.2 attempts to summarize some characteristics of the technical and human elements of organizational resources.

Concepts can be defined as the constituents of thought (Margolis and Laurence 2014). If the imprecise and uncertain elements of organizational resources are ignored in the management process, management would be a partly blindfolded activity. The second organizational resource question is detailed in Section 1.1.2.

1.1.2 What Are the Elements in the Organizational Resources?

The challenge for management is to understand and make use of all kinds of elements of organizational resources: from technical to human and abstract elements. Technical elements we can measure and utilize accordingly, but human and abstract elements are more challenging. Since we cannot really measure them, we cannot really easily utilize them for management purposes either.

According to Zadeh (2005), our perceptions of objects are vague and linguistic in nature. This means that management should somehow be able to make use of linguistic and vague perceptions of organizational resources for management purposes regardless of the type of element at hand. This is a fundamental prerequisite to really understand and make use of organizational resources for management purposes. If this prerequisite cannot be met, management will be a partly blindfolded activity. Luckily, there are approaches and methods that consider the uncertainty and imprecision that is present everywhere in the world. To be able to understand and utilize the whole spectrum of organizational resources for management purposes is very important, since it determines the basis for operational success, the basis for lifetime costs, and value creation. These are indeed the most important aspects of any organization. Therefore, organizational resources must also be systematically and continuously developed and managed, and at the same time use of them must be made for management purposes. This sounds like a continuous management cycle!

In many cases, wrong decisions have been made regarding organizational resources in a certain situation due to inadequate, biased, or just plain wrong information and clear violation of right decision principles (Table 1.1). Some aspects of the management of organizational resources are external and cannot be influenced or predicted (cf. Aldag and Stearns 1991; Ruohotie 2005). Since we cannot influence external factors, organizational resources should be dynamic and adaptive to changes in the environment.

In changing situations, managers sense and try to understand the organizational resources they are managing in their mind according to their experience. But it is practically impossible to get a reliable grasp of the concepts, their relationships, and indicative evidences. Many important concepts, their relationships, and indicative evidences that are related to organizational resources are vaguely understood and recalled in changing real-life situations. The reason for this is that organizational resources are very complex management objects that have multiple technical as well as human and abstract elements actively present in management situations simultaneously. Each element can have several different viewpoints that can be considered. Some of these viewpoints may be clearer than others. Some viewpoints we can observe and measure, but some others we can neither observe nor measure. Some viewpoints of organizational resources we are not even aware of. Perhaps, a real expert of some work domain is aware of all the elements of a certain organizational resource and also of all different viewpoints on these elements. But there is no such expert in the whole world who is aware of all the elements of organizational resources and all of their different viewpoints in changing contexts. Most people who are working with organizational resources are not domain experts, for example, new employees.

Many solutions typically fail due to the lack of holism (Jackson 2004). This means that decisions should be based on real situational knowledge of the whole object rather than knowledge of individual elements of the object, and instead of "educated guesses," intuitive feelings or information that represents only some aspects of an organizational resource. In a way, managers in many cases are forced to work and make decisions with partial information and in violation of right decision principles as shown in Table 1.1. The third organizational resource question is described in Section 1.1.3.

1.1.3 How to Make Use of All Kinds of Organizational Resources for Management Purposes?

Let us assume that all kinds of organizational resources are made somehow available for managers. Then, a problem arises from the complexity of these resources since people are not very good at making precise yet significant observations and statements about a system's behavior as the complexity of the system increases (Zadeh 1973). In other words, people cannot really understand complex systems or their behavior, such as organizational resources. It is common sense that something we cannot understand, we cannot manage. This leads to a need to clarify all the elements of organizational resources in order to help people to understand these complex systems.

The reality facing managers is so complex and fast-changing that accurate mathematical modeling really is impossible (Jackson 2004). For example, an accurate mathematical model of a human or a human system is not a realistic idea. We can keep in mind that humans are integral parts of many organizational resources. In order to be able to understand organizational resources in specific situations and contexts, some kind of modeling is clearly needed. Otherwise, required structures are not available in the decision-making situation and decisions cannot be based on real holistic knowledge. It seems that managers would need support that is beyond the capabilities of traditional decision support systems to be able to deal with the complex nature of organizational resources. Now, we can pose the fourth organizational resource question as mentioned in Section 1.1.4.

1.1.4 How to Handle the Complexity of Organizational Resources for Management Purposes?

There are always multiple internal and external stakeholders to organizational resources. This means that there must also be multiple perceptions of organizational resources as well. There are at least individual and work-role-based differences in perceptions. Most organizational resources are of a subjective nature and therefore stakeholders and individuals perceive them differently from their own viewpoints. One of management's greatest challenges arises from the natural tension between different types of people working together in an organization (Casciaro and Lobo 2005). The perception of coworkers (extrospection) often differs from the individual's self-perception (introspection). In other words, extrospection and introspection regarding the same organizational resource may result in totally different perceptions. Different stakeholders also have different interests in organizational resources. Management decisions regarding organizational resources are also perceived with varying levels of understanding and enthusiasm. Stakeholders' and individuals' perception or organizational resources also change according to time and situation. Dynamic subjectivism is therefore one typical quality of organizational resources. The meaning of an organizational resource to an individual is based on an individual's accumulated experience in life. Thus, it seems very likely that any two people will perceive the same organizational resource differently. We cannot change that, but maybe we can make organizational resources easier to perceive and easier to communicate about.

In addition to perceiving the existing reality of organizational resources, stakeholders and individuals have their own aspirations, visions, and goals regarding the resources. It is rare that precise objectives have been defined on which all stakeholders can agree (Ackoff 1986; Jackson 2004). Multiobjectiveness is typically hidden from others, including the management. This can mean trouble for managers. This is especially problematic in goal setting and during change processes. Hidden multiobjectiveness cannot really be controlled or managed. The lack of transparency provides space for unexpected surprises in the organization. Organizational resistance to organizational resources may also indicate hidden and conflicting objectives. Organizational resistance is considered one permanent factor that management has to consider. It would be better if multiple objectives were transparent. Then, it could be possible to integrate different transparent objectives for the benefit of individuals and whole organizations. By doing that, it may become possible to be proactive toward foreseeable internal and external events before these events actually take place. If we want to achieve that, we need to enable some kind of transparent future view with regard to organizational resources. The fifth organizational resource question is explained in Section 1.1.5.

1.1.5 How to Have Transparent Current and Future Views on Organizational Resources?

The fundamental management challenge can be formulated as follows: how can the management manage organizational resources adequately when these objects are subject to many different kinds of challenges as described in this chapter? What kind of management framework should we build to help managers in their difficult task? How can a new type of management paradigm be created with which clarified organizational resources can be transparently and better managed? Usually, management has the clear picture of only some part of organizational resource(s) with regard to the challenges described earlier. As a result, management decision making is often done in violation of right decision principles (Table 1.1). Table 1.3 summarizes different kinds of challenging questions facing organizational resource management. The goal is to offer practical answers to these questions in this book.

TABLE 1.3
Organizational Resource Questions

What are our organizational resources today?
What are the elements of the organizational resources?
How to make use of all kinds of organizational resources for management purposes?
How to handle the complexity of organizational resources for management purposes?
How to have transparent current and future views on organizational resources?

The next sections formulate propositions and a framework to answer to these organizational resource questions. First, the original Ingenious Management propositions are introduced (Kantola 2005) in Section 1.2, and after that the resource management propositions are introduced in Section 1.3.

1.2 INGENIOUS MANAGEMENT PROPOSITIONS

Research propositions are statements about the concepts presented in research that may be judged as true or false (Cooper 2002). Propositions are the shareable objects of attitudes and the primary bearers of truth and falsity (McGrath 2014). I do not want to go too far into the field of philosophy, and therefore the following simple definition of a proposition is adopted in this book: *A statement that can be judged as true or false.* The propositions for the management of organizational resources are defined and the framework they form together is presented. In Chapter 2 of this book, these propositions are explored further.

Five research propositions were created in a doctoral thesis, and they were named the propositions of Ingenious Management (Kantola 2005), as shown in Table 1.4. The propositions defined the structure of the thesis since they formalized the research problem and determined what the rest of the thesis should include. The propositions set out in the thesis required that proof was found and shown to be adequate to support them. Together, these propositions form the framework for Ingenious Management that is illustrated in Figure 1.3.

Figure 1.3 shows that all the propositions need to be true to meet the goal. In the thesis, all propositions were said to be true based on empirical evidence (Kantola 2005). Figure 1.4 shows the functional model of the propositions and can be read as follows: people in an organization perceive MOs (management objects) in their conscious experience. These MOs are specified as ontologies and become MOOs (management object ontologies). Some of them have a specified system boundary while some do not. The ones with a system boundary are capable of producing meta-knowledge. People evaluate both types of MOOs linguistically and form instances with the help of the Internet-based computing technology Evolute. Managers use

TABLE 1.4
The Ingenious Management Propositions

P1: Considering organizational resources as management objects and constructing them as ontologies enables the current and future positioning of these resources.

P2: Two types of ontologies can be specified from the concepts of management objects:

- P2a: The ontology of a management object
- P2b: The ontology of a system (management object as a system)

P2a and P2b are management object ontologies.

P3: Global access to management object ontologies can be realized through a repository and interpreter.

P4: Instances (individuals' perceptions of management object ontologies) by introspection or extrospection are unique.

P5: Utilizing instances enables the Ingenious Management of organizational resources.

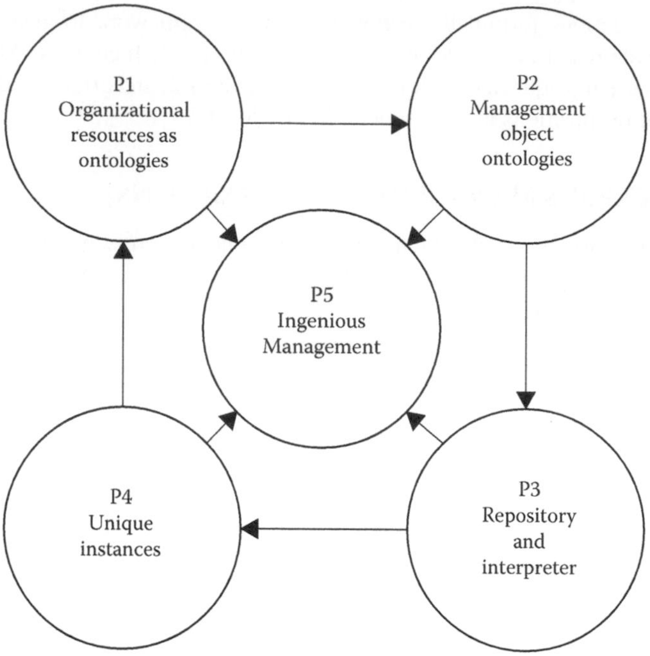

FIGURE 1.3 Framework of Ingenious Management. (From Kantola, J., Ingenious management, Doctoral thesis, Tampere University of Technology, Pori, Finland, 2005.)

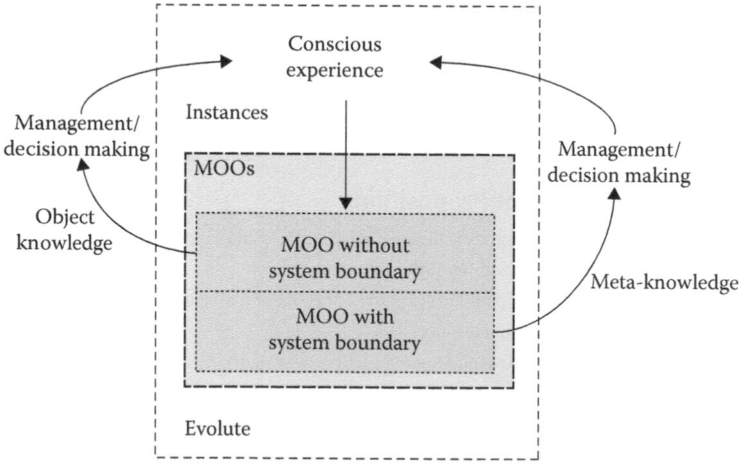

FIGURE 1.4 Functional model of Ingenious Management. (From Kantola, J., Ingenious management, Doctoral thesis, Tampere University of Technology, Pori, Finland, 2005.)

the visually presented results of instance-based reasoning to justify decisions and actions regarding MOs. This is repeated in periodic cycles. The approach is neutral to the content of MOOs (Kantola 2005), as shown in Figure 1.4.

For the purpose of this book, the rather academic Ingenious Management framework is evolved into the (hopefully) more practical and simple organizational resource management framework with four propositions that are described in Section 1.3.

1.3 RESOURCE MANAGEMENT PROPOSITIONS

Based on the previously described management challenges, the author proposes the following framework for the management of organizational resources. The framework consists of four propositions that all *need to come true in the right order in the organization* if we want to meet and overcome the management challenges described in the previous sections. Together, these propositions form a framework for management decision making and the basis for the actions that are planned following management decisions. The propositions to manage organizational resources are presented in Table 1.5.

Proposition 1 is based on the recognition of the resources that the organization utilizes to create value. It says that the organizational resources have to be comprehensively modeled. As we found earlier in this chapter, without comprehensive modeling of the objects we work with it is impossible to understand them in a holistic manner.

Proposition 2 is based on Proposition 1, and it says that these holistic models have to be brought to people so that their expertise can be utilized for the management efforts. People who are working in different work roles have front-row seats with the best close-up view to know what is going on with the resources. What goes well and what does not? The management office may have back-row seats and the distance from the actual reality of resources in different work and business situations may be

TABLE 1.5
Organizational Resource Management Propositions

Proposition	Explanation
1. Model relevant content	Modeling the conceptual content and structure of relevant organizational resources provides a holistic base to manage these resources.
2. Involve right people	Providing the models of resources (Proposition 1) to stakeholders using familiar language allows the best hands-on knowledge of resources to be included in management efforts.
3. Compute the meaning	Stakeholders' perceptions (Proposition 2) can be computed to give answers/meaning to the current reality and future target states of the resources.
4. Manage in the context	Utilizing the computed meanings of resources (Proposition 3) over time enables the management of these resources in their timely and specific context.

too great. There are evidences and indicators of organizational resources that are easily perceivable from the front row, but not from the back row.

Proposition 3 says that the perceptions of stakeholders from their own angles and perspectives (Proposition 2) can be transformed and shown within a holistic model of organizational resources (Proposition 1) that acts as the base for decision making by the management. Why? Because the holistic model according to Proposition 1 shows the big picture and it should be clear that management decisions should be based on the holistic view. Such a transformation should likely be computed by a computer, not by humans. Humans are not good at number-crunching. So-called "human-computing" (Law and Ahn 2011) and context-aware computing (Dey 2001) will have larger roles in the management part, where contextual and situational understanding play crucial roles.

Proposition 4 is based on Proposition 3. It says that when these computed meanings are "collected" systematically from stakeholders, this credible knowledge input can be used for the management of organizational resources. When this is done over time, periodically, the management can get a reliable picture of the current reality and future vision of their resources according to stakeholders. This kind of knowledge input is as reliable as human perceptions allow. The four organizational resource management propositions describe a hybrid human and computation management process. The computed meaning of organizational resources must be understood in a situational and contextual manner. This kind of context-aware human computing in management is something that humans can do much better than computers—at least in 2015. That is why humans still make decisions, not computers. Both have their important role in the management process according to these four propositions. Figure 1.5 illustrates how these propositions are connected to each other.

Each layer in the presented framework has a specific role in the whole and each layer enables the next layer to come true, and therefore each layer is needed for the framework to work in real life. Figure 1.6 illustrates this and can be read as follows: modeling enables involving; involving enables computing; and computing enables management. In other words, modeling, involving, and computing enable management when done in the right order. If any of the propositions is not true in the organization, the management does not use the best available explicit and tacit knowledge for the management decision making and thus violates the right decision-making principles in Table 1.1. This is a risky business for the organization! Figure 1.6 emphasizes the flow of these steps. According to Figure 1.6, any proposition that is false in the organization leads to unknown territory for management, and that is not a good place to be for an organization. Nobody wants unknown management or the management of the unknown!

This book attempts to show that modeling organizational resources is the first step to helping managers in their very difficult tasks. By systematically constructing organizational resource models, and systematically involving stakeholders' expertise in using the models within an organization, the degree of right decisions can be increased (Figure 1.1). Computing systems to show the meaning of stakeholders' perceptions of the management domain is necessary. Only basing management decisions on feelings, intuitive thoughts, and educated guesses is not good enough.

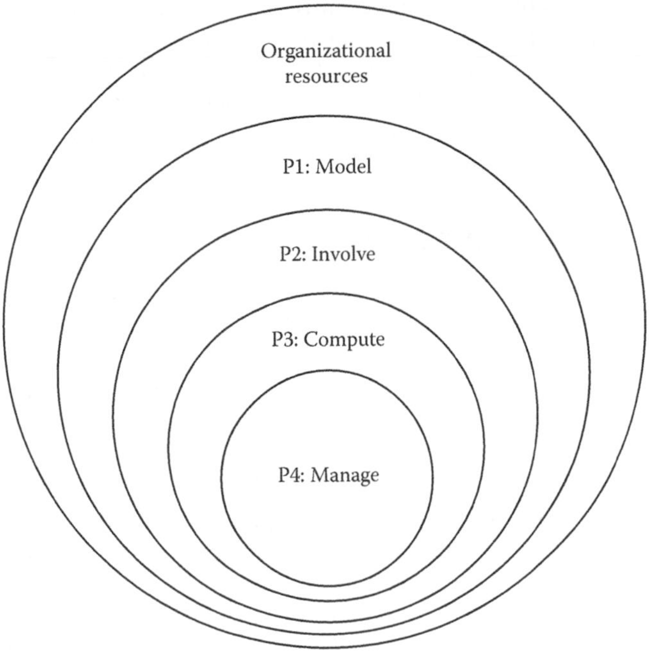

FIGURE 1.5 Propositions of organizational resource management form a layered framework.

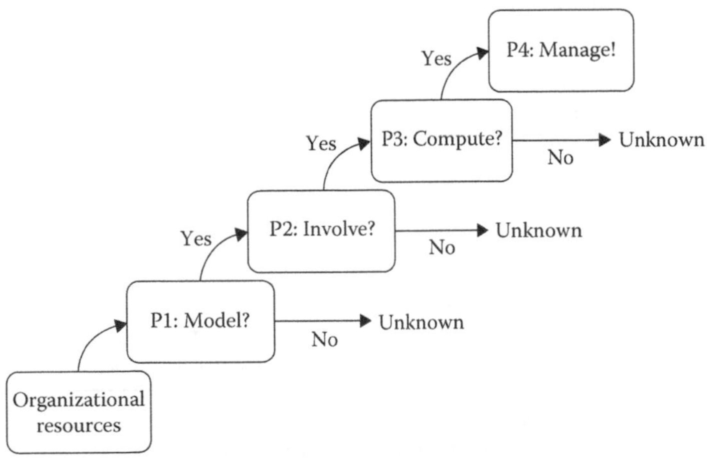

FIGURE 1.6 Each proposition must be true in the organization in order to avoid (1) unknown management and (2) the management of the unknown.

The reality is far too complex. Comprehensive modeling also ensures that the members of an organization "talk the same language" when current situations, development targets, and actions are being selected. When decisions direct development efforts and resources to the right targets, clear benefits are achieved. In fact, if true, this approach should be the only way to manage an organization. Section 1.4 describes the structure of this book.

1.4 STRUCTURE OF THE BOOK

This book is organized according to the framework presented in Figure 1.5. The book consists of 10 chapters that are shown in Figure 1.7. The first chapter introduced the topic by describing various aspects of organizational resources, and proposed a framework for the management of organizational resources based on earlier academic research. Chapter 2 describes the field of each proposition of the new proposed framework on a general level. Chapter 3 describes the Evolute approach as one solution for the management framework proposed in Chapter 1. Chapters 4 through 7 describe different tools that have been developed based on the Evolute approach as well as real-world cases where these tools have been used. Chapter 8 envisions how so-called knowledge increments can be used for innovation regarding organizational resources. In Chapter 9, some other existing approaches and methodologies that are being used for the management of organizational resources in organizations are overviewed. Chapter 10 discusses overall findings in this book and also presents potential future steps.

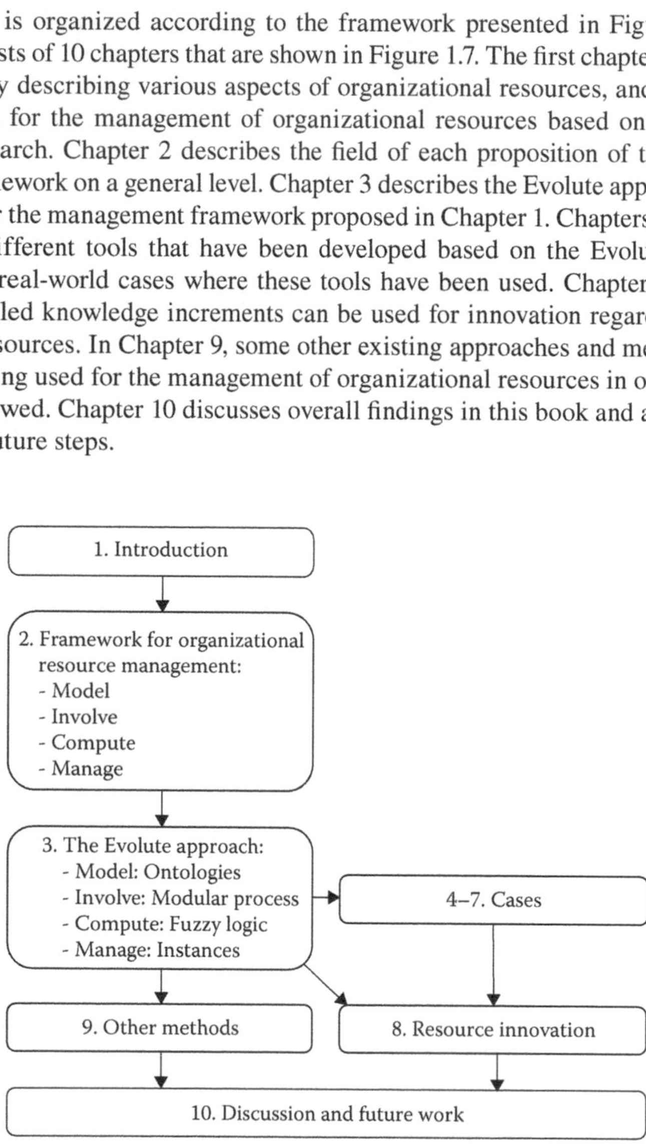

FIGURE 1.7 Structure and flow of the book.

REFERENCES

Ackoff, R.L. 1986. *Management in Small Doses*. New York: John Wiley & Sons Inc.

Aldag, R.J. and Timothy, S. 1991. *Management*. Cincinnati, OH: South-Western Educational Publishing.

Argyris, C. 1999. *On Organizational Learning*, 2nd edn. Oxford, U.K.: Blackwell Publishing.

Berry, J. June 2005. *Tangible Strategies for Intangible Assets*. New York: McGraw-Hill.

Casciaro, T. and Lobo, M. 2005. Competent jerks, lovable fools and the formation of social networks. *Harvard Business Review* 83: 92–99.

Cooper, D.R. and Schindler, P.S. 2002. *Business Research Methods*. Boston, MA: Irwin McGraw Hill.

Darwin, C.R. 1869. *On the Origin of Species by Means of Natural Selection, or the Preservation of Favoured Races in the Struggle for Life*, 5th edn. London, U.K.: John Murray.

Dey, A. 2001. Understanding and using context. *Personal and Ubiquitous Computing* 5(1): 4–7.

Jackson, M.C. 2004. *Systems Thinking: Creative Holism for Managers*. West Sussex, U.K.: John Wiley & Sons Ltd.

Kantola, J. 2005. Ingenious management. Doctoral thesis, Pori, Finland: Tampere University of Technology.

Kantola, J. and Karwowski, W. 2012. *Knowledge Service Engineering Handbook* (*Ergonomics Design and Management: Theory and Applications*). Boca Raton, FL: CRC Press.

Law, E. and Luis, A. 2011. Defining (human) computation. *Synthesis Lectures on Artificial Intelligence and Machine Learning* 5(3): 1–121.

Margolis, E. and Laurence, S. 2014. Concepts. In Zalta, E.N. (ed.), *The Stanford Encyclopedia of Philosophy*, Spring 2014 edition. Stanford, CA: Stanford University.

McGrath, M. 2014. Propositions. In Zalta, E.N. (ed.), *The Stanford Encyclopedia of Philosophy*, Spring 2014 edition. Stanford, CA: Stanford University.

Nonaka, I. and Takeuchi, H. February 1995. *The Knowledge-Creating Company: How Japanese Companies Create the Dynamics of Innovation*. New York: Oxford University Press.

Nonaka, I., Toyama, E., and Konno, N. 2000. SECI, Ba and leadership: A unified model of dynamic knowledge creation. *Long Range Planning* 33(1): 5–34.

Ruohotie, P. 2005. *Oppiminen Ja Ammatillinen Kasvu*, vol. 3. Helsinki, Finland: WSOY.

Suh, N.P. 1990. *The Principles of Design*. New York: Oxford University Press.

Suh, N.P. 2001. *Axiomatic Design: Advances and Applications*. New York: Oxford University Press.

Vanharanta, H., Magnusson, C., Ingman, K., Holmbom, A., and Kantola, J. 2012. Strategic knowledge services. In Kantola, J. and Karwowski, W. (eds.), *Knowledge Service Engineering Handbook*, pp. 527–555. Boca Raton, FL: CRC Press.

Zadeh, L. 1973. Outline of a new approach to the analysis of complex systems and decision processes. *IEEE Transactions on Systems, Man, and Cybernetics* 1(1): 28–44.

Zadeh, L. 2005. Toward a generalized theory of uncertainty (GTU)—An outline. *Information Sciences* 172(1–2), June 9: 1–40.

2 Framework for Organizational Resource Management

In this chapter, the organizational resource management framework is explained further. The framework is discussed in the following sections in order to find out what is required for successful solutions in each of the four parts in the framework.

2.1 PROPOSITION 1: MODEL RELEVANT CONTENT

Based on the introduction, we know that we need to model organizational resources somehow in order to "enable holism" for managers. Relying only on human capabilities is not enough. The first part of the organizational resource management framework examines different optional ways to model organizational resources. Without modeling, it is impossible to know what organizational resources consist of. Modeling organizational resources can be done from an object domain perspective and also from a system domain perspective. The object view brings object domain knowledge in management practice and examines how an organizational resource functions as an object. The system view adds meta-knowledge to management practice and examines how an organizational resource works as a system according to system parts and their interactions, the system functions, and the system border.

The content and structure of organizational resources are not comprehensively known by people in organizations. Because of this, an assumption is made that some kind of modeling of organizational resources is necessary. Modeling is a systematic description of an object or phenomenon that shares important characteristics with the object or phenomenon (Science Dictionary 2015). In this book, the role of modeling is to describe organizational resources in some way to show their content and structure to people and thus enable their full utilization in an organization. The "audience" for such representation is the stakeholders of organizational resources. These include internal and external stakeholders, that is, those individuals and groups of people who have some interest and role with regard to the resources, such as executives, project managers, salespeople, plant workers, customers, and suppliers. These people are then the users of the models of organizational resources in the course of their work. The models must support the filling of the existing knowledge gap in the content and structure of organizational resources in organizations.

2.1.1 Objects or Systems?

Organizational resources have a systemic nature. A system is like a group of interacting, interrelated, or interdependent components that form a complex and unified whole that has several essential characteristics (Anderson and Johnson 1997; O'Connor and McDermott 1997). In an organization, the interactive tangible and intangible resources and their parts together form systems. Some of these systems cross organizational boundaries and can become elements in a larger system of systems (Marek et al. 2014). There are a few basic types of system: mechanical/electrical systems, human systems, electronic/telecommunications systems, ecological systems, and biological systems, and their combinations, as well as other kinds of manmade systems (Haines 2000). Organizational resources can belong to any of these types. Another way of looking at systems is in terms of their openness. A closed system is isolated from its environment, while an open system interacts with its environment, accepting inputs and generating outputs (Haines 2000). Organizational resources are clearly open systems that interact with their environment.

Should we view organizational resources as objects or systems? In Figure 1.2, there are organizational resources, such as people, technology, and processes. We can observe these resources as objects in an object domain. We can also view these same resources as systems in a system domain. We can see that people, technology, processes, etc., are made of integral, interconnected parts that need each other for the whole system to function and to achieve its goals. We can zoom out and see the whole organization as a system or as a system of systems (Marek et al. 2014). We can zoom in and see smaller systems inside bigger systems. All organizational resources are systems with many interconnected elements in them. Depending on how we choose, we can have different and complementary ways of viewing organizational resources, as shown in Figure 2.1.

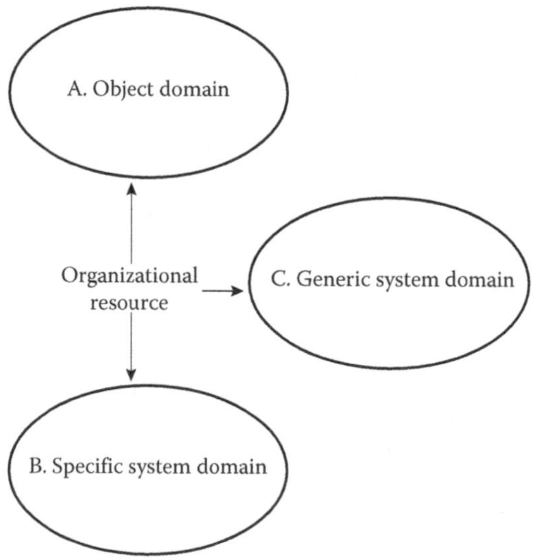

FIGURE 2.1 Organizational resources can be viewed in different and complementary ways.

From the modeling point of view, how we choose to view organizational resources will become the basis for modeling the resources. Figure 2.2 shows different (A) object-, (B) specific system–, and (C) generic system–based views through which to model organizational resources.

Type A in Figure 2.2 answers the question: What is the organizational resource like as an object? Type B answers the question: What is the organizational resource like as a specific domain system? Type C answers the question: What is the organizational resource like as a generic domain system? Combinations A–B, A–C, B–C, and A–B–C answer the combinations of these questions. An example of an object domain (A) is a production line or human competency in a workshop such that systemic interaction between different elements is not embedded in the model. An example of a specific domain system (B) is a process in an office and/or human competency where system dynamics, that is, interaction and the causality of systems and their elements, are embedded in the model. Two examples of a generic domain system (C) are Miller's living system theory (Miller 1978) and Samuelson's InformatiCom concept of how living systems behave (Samuelson 1981), that is, they accept information, matter, and energy as inputs, and through subsystem functions (which communicate, control, support, and produce), information, matter, and energy are generated as outputs.

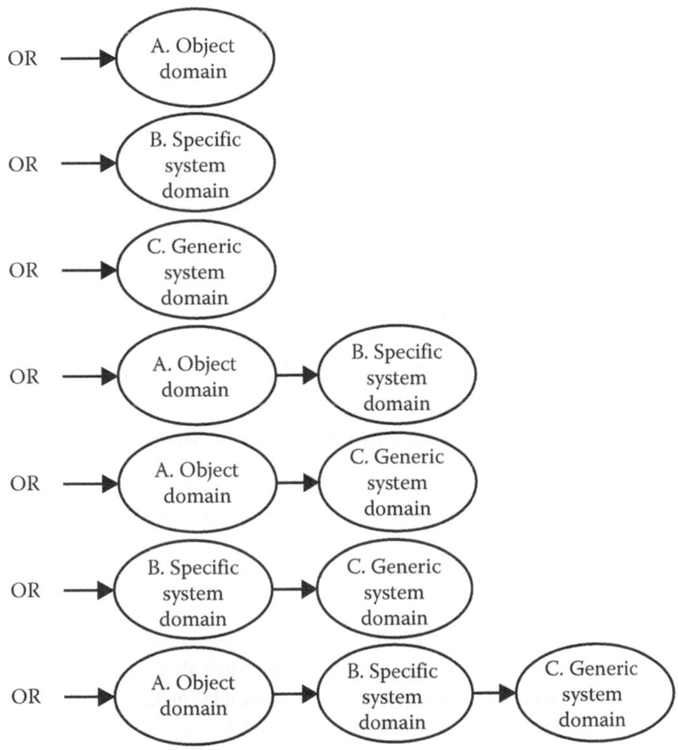

FIGURE 2.2 Optional basis to model organizational resources (OR).

Organizational resources are clearly systems in addition to being objects. The modeler can choose whether to embed the interaction of system parts in the model, or leave it out to be considered by the model user. These different views are important basic blueprints for developing complementary models of organizational resources that support stakeholders' full involvement. Such complementary models together provide better understanding of the current reality and the future direction of organizational resources. The nature of a specific organizational resource gives an indication as to which base type model can be developed. Another factor in determining the base type is the intended use of the model. Models are developed for some purpose and their use has aims in organizations. One more factor that influences the base type is the constraint of knowledge availability. Do model developers and users have access to the adequate knowledge that is needed for the chosen model base type?

A system boundary defines which elements belong to the system and which elements do not. Through an open system boundary information, energy and matter go into the system and come out of the system (Luhmann 1995; Miller 1978; Samuelson 1981). It is useful to try to define where the system boundaries of organizational resources are. Then, for example, metrics for measurement, control, and action can be specified on the boundary. However, it is not necessarily an easy task to define the system boundary. Table 2.1 summarizes the differences of A–B–C views on model organizational resources.

The outcome and performance of organizational resources can be observed either as an object or a system against set goals. Organizations typically set goals for the objects and for the system and measure whether the goals were achieved. But is this enough? Can we not foresee and predict the outcome of objects and systems without the need to wait and see what comes true? Can we not have a more meaningful view of our organizational resources in addition to the numbers always used by the management, such as KPIs (key performance indicators)? In Section 2.1.2, we look at different types of model and explore what types of model are suitable for modeling organizational resources.

TABLE 2.1
Different Viewpoints on Organizational Resource Modeling

	Object Domain View (A)	Specific Domain System View (B)	Generic Domain System View (C)
Domain concepts embedded in the model	Yes	Yes	No. generic system concepts, yes
The relationship of domain concepts	Concepts can be grouped, clustered, and specified in hierarchy. Causalities to be interpreted by the model users	Causalities (direction and strength) are embedded in the model	No
System boundary	No	Can be specified	Yes

2.1.2 Types of Models

There are many different types of models that are used in an organization: physical models, mathematical models, structural models, software models, business models, causal models, data models, economic models, ecosystem models, graphical models, macroeconomic models, psychological models, statistical models, system models, toy models, metaphysical models, conceptual model, semantics models, logical models, scientific models, architecture models, linguistic models, etc. Different model types are of course used for different purposes. In this book, a model type that can adequately capture the nature and characteristics of organizational resources must have the ability to

- Describe tangible, such as technological, elements labeled with precision and certainty
- Describe intangible, such as human, elements labeled with imprecision and uncertainty
- Describe relationships of tangible and intangible elements of organizational resources labeled with ambiguity and a degree of interaction

Table 2.2 lists those model types that might be usable for modeling organizational resources: mathematical, conceptual, system, and causal models. Mathematical models describe the object using mathematical language. Conceptual models contain concepts to describe the object. System models are multiview descriptions of a system. Table 2.2 roughly maps the suitability of each type of model to the requirements listed earlier.

Precise mathematical models of humans and human systems are impossible to develop as well as to maintain. On the other hand, conceptual/semantic models and system models seem to be better suited to modeling organizational resources even though they cannot precisely model all elements. Systems models, in turn, focus on describing the interaction between elements, whether tangible or intangible, but not on the content and structure of the elements.

TABLE 2.2
Suitability of Model Types for the Modeling of Organizational Resources

	Ability to Describe		
Model Type	**Tangible Elements/Content**	**Intangible Elements/Content**	**Relationships between Elements/ Systemic Structure and Behavior**
Mathematical models	X		
Conceptual/semantic models		X	X
System models		(x)	X
Causal models			X

Another way to look at different types of model is their flexibility. Static models do not change. They provide a snapshot of an object. Comparative models allow the observation of an object for a longer period of time and enable a comparison to some static model or state. Dynamic models are in interaction with the environment and allow observation of how to change the object of observation and how to develop the model itself (McGarvey and Hannon 2004). Clearly, dynamic models of organizational resources are the ones that organizations want to have, since they support the development of an organization in interaction with its environment through developing the resources of the organization. At the same time, the models of organizational resources are evolving themselves. Stakeholders can be involved in the management through organizational resource models. Organizational resource models have their lifecycle and they evolve as part of the process. Next, we look into the second part of the framework, which aims to involve people.

2.2 PROPOSITION 2: INVOLVE PEOPLE

In this section, we examine how to present organizational resources to relevant people in formats that are familiar to them and how to involve these people in management action. Relevant people include employees, managers, and all other internal and external stakeholders who have some connection or interest, currently or in the future, with regard to the organizational resources. Presentation is a required step since that is the only way we can enable stakeholders to participate in the management. We cannot really involve people secretly without showing the model to them in some way. Presentation is very important since the best understanding and expertise is usually located near the resources, whether this be on the factory floor or some other place. If we do not present organizational resource models to stakeholders, their expertise remains beyond the reach of management. Involving the best understanding and expertise should be in the interest of all organizations and their managers. But what is a good enough way to present organizational resources to all relevant stakeholders?

In order to get the full benefit of using organizational resource models in management, it is important to involve all relevant stakeholders to their full potential in each case. There is a challenge in presenting the models to people since the terminology may not be familiar to people who are doing their work. The presentation of models describing organizations in action is something that is typically not easy for people to understand. This means that organizational resource models need to be presented using such language as is familiar to people in different work roles. There is a need to present organizational resource models using layman's terms to stakeholders. Another point to note is that we likely cannot observe intangible or even tangible resources directly, but instead we can observe signs and indicators of those in our working environment.

In this book, involving people refers to taking those people who have relevant knowledge of organizational resources in the management process and engaging them in decision-making and resulting action processes. It is common sense that adding more relevant knowledge of the object at hand to the decision-making process will likely improve the quality of the decisions regarding the object. Inadequate knowledge input to the decision-making and management process cannot yield excellent results. Relevant knowledge refers to the expertise and understanding of the object and its behavior that various individuals and stakeholders have. Put simply, the goal is to have the most credible human knowledge input for the management process as possible. This is the first step in involving people. The second step is to involve the people in the decision-making process regarding the resources since they understand the resources in their current and future contexts. It is also likely that they are motivated in participating in the decision-making process since they are already involved in providing their knowledge input to the management process with the help of organizational resource models.

However, it is not enough to include only the human knowledge from relevant stakeholders in the management process. Any earlier knowledge that can give guidance should also be included. Any relevant material, such as project reports, books, and articles, can be included in the process as well. Finally, it is very important to include the management's understanding of the organization in its current state and context as well as in its envisioned future state and context, that is, the vision, mission, and strategy of an organization. Hand in hand with the strategy go constraints, that is, what is realistic and what is not possible time-wise, money-wise, etc. These three main knowledge elements of organizational resources are must-haves in the management decision-making process in any organization: (1) relevant tacit (human) knowledge from stakeholders with the help of organizational resource models, (2) the organization's strategy, and (3) other relevant knowledge. Section 2.3 examines the third part of the framework, which is computing.

2.3 PROPOSITION 3: COMPUTE MEANING

In the third part of the framework, the stakeholders' perceptions of organizational resources are captured and used. Stakeholders' perceptions of the current reality and envisioned future of organizational resources in their contexts is the knowledge input to the management process. Collecting these perceptions enables the management of organizational resources. However, captured perceptions as such need further processing and conversion, so that they become usable for the management. Some kind of computing of the meaning of stakeholders' perceptions to the decision-making process is necessary. We need to be able to reason based on the collective understanding and expertise that people have with regard to organizational resources. After we have involved relevant stakeholders, we need to use some technology and methods to convert the understanding and expertise that stakeholders have relating to management decision making. With computing algorithms we can imitate a human style of expert reasoning based on captured perceptions. The usual method of reasoning in organizations is to rely on management's expertise in different situations. But in this book, we need to have a reasoning entity that is capable of transforming stakeholders' perceptions into immediate and meaningful knowledge inputs for the management. This kind of reasoning entity enables the full participation of all stakeholders in the management of organizational resources.

Therefore, for management purposes the models of organizational resources must be computable, that is, meaningful outcomes can be computed based on the models. There can be three different types of input streams for these models (Kantola and Karwowski 2012; Vanharanta et al. 2012):

1. *Data streams*: For example, from sensors or numerical outputs from algorithms (explicit)
2. *Information streams*: For example, online text and visual material (explicit)
3. *Human knowledge streams*: For example, the expertise that people in different work roles have about their work and working environment as well as situational understanding of the organizational resources in specific contexts (tacit)

We can see that collective human expert knowledge input is crucial for management purposes. Decisions concerning organizational resources without such human knowledge inputs are risky, of course. Next, we look at the fourth and final part of the framework, which is management.

2.4 PROPOSITION 4: MANAGE IN THE CONTEXT

In the fourth part of the framework, we look at how the management of organizational resources should be organized based on data, information, and human knowledge inputs for the management decision-making process. Organizational resources are changing, their models are evolving, their contexts and environment are changing, and the perceptions of stakeholders are changing. Repeating the utilization of the collection of perceptions in frequent cycles in an organization is clearly needed. Otherwise, the ever-continuing dynamics of internal and external environments would be ignored. The meaning of perceptions is combined with the company's strategy and other relevant explicit knowledge related to the organizational resources. In this way, the management decisions are always based on validated resource models and collective and situational expertise. This means that the management decisions that will be taken are well justified.

The stakeholders and the models of organizational resources have an integral role in a continuous management process (Figure 2.3):

1. Utilize expertise to make organizational resource models.
2. Utilize the models for decision making.
3. Implement decisions, that is, take action.
4. Evaluate the successfulness of implementation.
5. Learning and increased expertise.

Figure 2.3 shows that the next step is always based on the previous step, and as the process goes on the expertise and the quality of the models will increase. At the same time, the degree of rightness of decisions increases and a bigger portion of the right decision space can be filled, as shown in Figure 1.1. In Chapter 3, one solution for the framework is presented. It is named the Evolute approach.

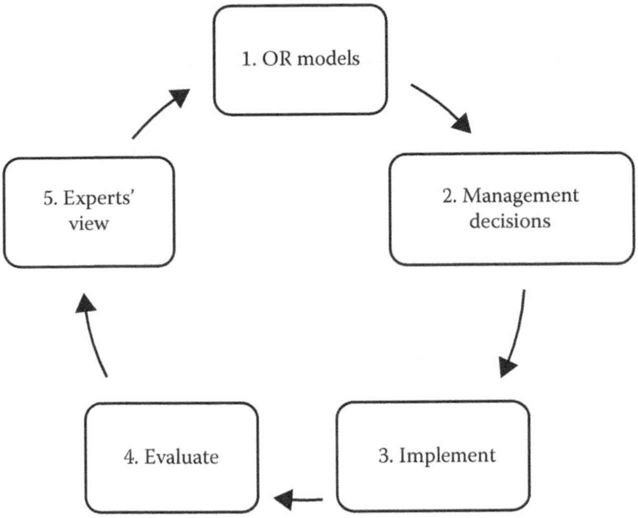

FIGURE 2.3 Stakeholder and organizational resource models are integral parts of a continuous management process.

REFERENCES

Anderson, V. and Johnson, L. 1997. *Systems Thinking Basics: From Concepts to Causal Loops*. Waltham, MA: Pegasus Communications, Inc.

Haines, S.G. 2000. *The Complete Guide to Systems & Systems Thinking and Learning*, 1st edn. Amherst, MA: HRD Press.

Kantola, J. and Karwowski, W. (eds.). 2012. *Knowledge Service Engineering Handbook (Ergonomics Design and Management: Theory and Applications)*. Boca Raton, FL: CRC Press.

Luhmann, N. 1995. *Social Systems*. Stanford, CA: Stanford University Press.

Marek, T., Karwowski, W., Frankowicz, M., Kantola, J., and Zgaga, P. (eds.). 2014. *Human Factors of a Global Society: A System of Systems Perspective*, 1st edn. Boca Raton, FL: CRC Press.

McGarvey, B. and Hannon, B. 2004. *Dynamic Modeling for Business Management: An Introduction*. New York: Springer.

Miller, J.G. 1978. *Living Systems*. New York: McGraw-Hill.

O'Connor, J. and McDermott, I. 1997. *The Art of Systems Thinking: Essential Skills for Creativity and Problem Solving*. London, U.K.: Thorsons.

Samuelson, K. 1981. In informationcom and multiway video communication as a cybernetic design and general systems technology. In *Proceedings of the SGSR 25th Annual Conference*, Toronto, Ontario, Canada.

Science Dictionary. 2015. Modeling. Dictionary.com. In *The American Heritage® Science Dictionary*. Houghton Mifflin Company. http://dictionary.reference.com/browse/modeling. Accessed April 7, 2015.

Vanharanta, H., Magnusson, C., Ingman, K., Holmbom, A., and Kantola, J. 2012. Strategic knowledge services. In Kantola, J. and Karwowski, W. (eds.), *Knowledge Service Engineering Handbook*, pp. 527–555. Boca Raton, FL: CRC Press.

3 Evolute Approach

The Evolute approach addresses all four organizational resource management propositions in a specific way. How this is done is explained in this chapter proposition by proposition. The approach emphasizes the importance of involving people in the management process with the help of specific computing solutions. The Evolute approach consists of a computing system and methodology that have been developed since the end of the 1990s.

The Evolute approach applies ontology engineering to model organizational resources according to Proposition 1. The Evolute approach follows a modular process where individuals and stakeholders are involved and their perception and understanding of organizational resources are sought and collected. This kind of knowledge input addresses Proposition 2. The Evolute system is a computing platform and a technology that computes and visualizes the meaning of the knowledge input collected from stakeholders. The Evolute system addresses Proposition 3. The computing in the Evolute system is based on soft-computing methods and algorithms in order to cope with imprecision and uncertainty embedded in natural language and human knowledge inputs. In particular, fuzzy sets and fuzzy logic (Zadeh 1965, 1973) are applied. Finally, Proposition 4 is addressed in the Evolute approach so that the management uses the computed current and future meaning of organizational resources in their real situational management context in the organization. Stakeholders can be involved in this management step according to the modular process (Table 3.1).

The Evolute approach attempts to meet the management challenge described in Chapter 1. The fundamental idea behind the approach is based on the precisiation of the meaning (i.e., making the meaning more precise and clear) of organizational resources in their real contexts for management purposes on three levels. Each level of precisiation is necessary to clarify to the management what the current reality of organizational resources is like and which elements require further attention from the management in the future. Clearly specified content is brought to current and future contexts. In this chapter, each step in the Evolute approach is described.

One of the biggest challenges that organizations are facing today is that there is too much explicit knowledge available and at the same time too little tacit knowledge available for the practical use of management. The explicit knowledge available is also too complex to be made sense of in management work. One attempt to address these problems is knowledge service engineering, which is an emerging field in the scientific and application worlds focusing on the joint systems of data networks, information networks, and human knowledge networks. The goal of knowledge service engineering is acquiring and utilizing data, information, and human knowledge to produce high-performance joint knowledge services to support the knowledge economy of the twenty-first century (Kantola and Karwowski 2012). The key is to

TABLE 3.1
Evolute Approach Addresses Each Proposition in a Specific Way

Organizational Resource Management Proposition	Evolute Approach
1. Model relevant content	Ontology engineering is applied to model the content and structure of organizational resources.
2. Involve people	Knowledge input is collected online from individuals and stakeholders.
3. Compute meaning	Fuzzy logic-based Evolute system computes the meaning of organizational resources today and in the future.
4. Manage in the context	Computed meaning is examined in the current and future contexts for decision making.

apply methods and tools that are capable of addressing the complexity and meaning of explicit and tacit knowledge in a practical manner. Sections 3.1 through 3.4 explain how the Evolute approach addresses each of the four propositions in the framework.

3.1 PROPOSITION 1: MODEL WITH ONTOLOGIES

By definition, semantic models and conceptual models are suitable for the modeling task described in this book. Semantics refers to the meanings of words and phrases in specific context. Semantic methods and models are ways to do this. Semantic models can be supported by ontologies and ontology engineering. In Section 3.1.1, we look at how ontologies are applied to model organizational resources according to the Evolute approach.

3.1.1 Ontologies

Ontologies sound promising and they might bring managers much needed support in their work. Typically, ontology is defined as the specification of the conceptualization of a domain (Gruber 1993). Conceptualization is the idea of (part of) the world that a person or a group of people can have (Gomez-Perez 2004), and ontology defines the common words and concepts (meanings) that are used to describe and represent an area of knowledge (Orbst 2003). Ontologies represent a method of formally expressing a shared understanding of information (Parry 2004). The main parts of ontologies are classes (concepts), relations (associations between the concepts in the domain), and instances (elements or individuals in ontology) (Gomez-Perez 2004). Ontologies represent knowledge in such a way that the content and structure of ontologies can be read and understood by machines (Uschold and Gruninger 1996). Therefore, ontologies also enable the computational processing of data, information, and knowledge.

It seems that ontologies can provide an approach to specify and manage organizational resources in a holistic way. In other words, without utilizing ontologies it is difficult for a manager to perceive, manage, and develop organizational resources in the right way since they cannot really follow the content and structure of their complex resources. Ontologies promise a base to share knowledge and thus enable common understanding of a domain that can be communicated between people and application systems (Davies et al. 2003). Constructing organizational resources as ontologies seems an attractive approach to position and manage these resources now and in the future. These domain ontologies are called organizational resource ontologies in this book.

3.1.2 Ontology Lifecycle

Ontology engineering refers to the ontology development process, ontology lifecycle, and the methodologies, tools, and languages for building ontologies (Gomez-Perez 2004). The ontology development process is similar to application development, sharing the major concepts of design, development, integration, validation, feedback, and interaction (Edgington et al. 2004). In the knowledge management context, an ontology development process has the following path (1) feasibility study (project setting), (2) ontology kickoff (specifications and baseline taxonomy), (3) refinement (knowledge elicitation with experts), (4) evaluation (using and testing), and (5) application/maintenance and evolution (organizational process) (Sure et al. 2004; Sure and Studer 2003). The last three steps are iterative. A basic requirement of feasibility is that there should be financial benefits to using ontologies.

The world is changing and our knowledge of the world evolves. Ontologies attempt to capture our knowledge of the changing world (Mitra and Wiederhold 2004). Therefore, ontology is not a static model of knowledge, and adjustment to ontology should be made continuously (Edgington et al. 2004). Organizational resource ontologies follow the normal parts of a lifecycle: birth, development, and death. Sometimes, they may also be split into parts that start to live their own lives. A number of ontologies may also be joined together to form a new, more comprehensive ontology.

Changes in ontologies can be caused by (1) changes in domain, (2) changes in conceptualization, or (3) changes in specification (Klein et al. 2003). The factors that influence the lifecycle of organizational resource ontologies include (1) the fact that better knowledge and understanding of organizational resources enables the development of better ontologies and (2) the fact that reflection on the changes in a specific business and operation context enables better adjustment and application of ontologies in the organization. Nobody can really predict how the lifecycles of organizational resource ontologies will evolve in the future. Therefore, in this regard, no fixed or permanent structures should be planned or built in organizations. Instead, agile and flexible frameworks, constructs, methods, and tools that can support the lifecycles of organizational resource ontologies dynamically in organizations should be preferred. This is a better approach for ontology developers and users as well.

3.1.3 Ontology Development in Organizations

Ontologies can be developed in an automated manner (Segev and Gal 2008; Segev and Quan 2012), in a human-centric manner or in some hybrid combination of the two. The Evolute approach is based on the human-centric manner. There are reasons for this. First, the domains of organizational resources are in large part conceptual. Therefore, human understanding is necessary to collect and organize the concepts. Second, many of the domains of organizational resources are conceptually new, and they have not been specified as ontologies that people could easily utilize in their organization. This means that there may not be much literature available. Putting together such new conceptual structures is something that computers cannot do yet in 2015. A lot of understanding of the conceptual world and its structures resides in the minds of people who are working with these organizational resources in different kinds of organizations. They have observed what is relevant and what is not; what works and what does not; what is cause and what is effect in practice; and so on. There are two kinds of relevant concepts: internal and external. Figure 3.1 illustrates these two types of concept.

According to the Evolute approach, the ontology development process is a human-centric exploratory process. The automated development of organizational resource ontologies is simply not possible due to the fact that tacit knowledge is needed. Typically, two kinds of people are needed in the ontology development process: ones who understand ontology engineering and ones who understand the domain of the organizational resource. These people can be researchers or practitioners. It is very likely that several people together can understand the domain better than one expert. Therefore, it is good to involve people with experience in the process in addition to explicit sources. There are many ways in which this development process can be organized in academia, organizations, and public, for example, in social media, as shown in Figure 3.2. The smaller circles are different types of organizations and the larger shapes represent different kinds of cooperation between these organizations, that is, who is cooperating and how.

Academic and/or private exploratory research is necessary regardless of how the process is organized. Ontologies can be built in research projects in one or more universities or organizations together. The outcomes of such research projects can be disseminated as books, reports, articles, theses, services, etc. Ontologies will develop over time as the domain evolves and as researchers and practitioners learn more about the domain and contexts. When the ontology evolves, new concepts can be added, modified, or deleted by someone. There must be a maintaining party who will take care of ontology lifecycle and maintenance. Ontologies are certain kinds

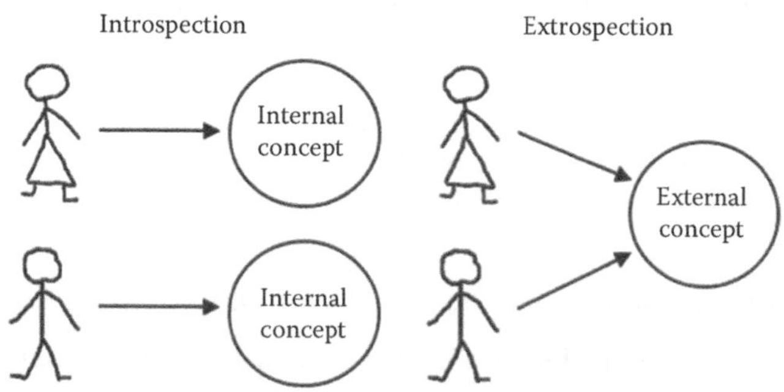

FIGURE 3.1 Internal and external concepts are observed through different channels.

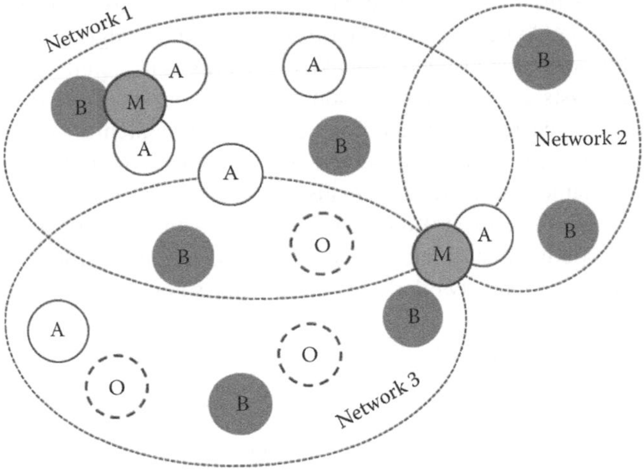

FIGURE 3.2 The ontology development process can be organized in many different ways (A, academic; B, business; O, organization; M, moderator).

of models of reality, and as conceptual reality changes the ontologies must change accordingly. When ontology development process is being planned and set up, it is necessary to consider ontology lifecycles as well.

3.1.4 Ontology Repository

Nonaka et al. (2000) write that management has to redefine the organization on the basis of the knowledge it has. An inventory of the knowledge assets is the basis for creating a strategy to build, maintain, and utilize the firm's knowledge assets effectively and efficiently. Firms need to "know what they know" even though it can be difficult to locate existing knowledge and get it to where it is needed (Davenport and Prusak 2000). Davies et al. (2003) wrote that there are three important aspects related to ontologies and the tools that support their use as part of the semantic web (Berners-Lee et al. 2001) of the future: (1) acquiring ontologies and linking them with large amounts of data, (2) storing ontologies and their instances, and (3) querying and browsing semantically rich information sources. The tools for supporting the use of organizational resource ontologies must also meet these requirements as described by Nonaka et al., Davenport and Prusak, and Davies et al. In order to utilize the knowledge stakeholders have of organizational resources for management purposes, a technology that enables ontology-based computing and reasoning is needed.

In order to store organizational resource ontologies in a computable format, a repository is needed. It could be the organization's own database or a cloud solution. In addition, the efficient and effective use of organizational resource ontologies clearly requires that all stakeholders of the relevant domains need to be able to access and use these knowledge assets. Today, this means that the repository of these ontologies must be available on the Internet/cloud or intranets. An online global access to organizational resource ontologies is actually required since organizations are global and functions can be globally dispersed. A repository containing organizational resource ontologies becomes the inventory of knowledge assets that an organization has, following the vision of Nonaka et al. This is indeed the most important asset! Table 3.2 lists organizational resource ontologies in the Evolute ontology repository with the current language translations. The content of the repository is evolving and today's status can be observed at http://www.evolutellc.com/tools.aspx. The developer's name and references are also listed.

In addition, there are several ontology researches in progress, for example, in the areas of knowledge management, new product development, logistics, out- and insourcing, value creation networks, and future work roles. Table 3.3 shows the conceptual structure of one sample ontology in the Evolute ontology repository. The cardioid ontology was developed by Professor Karwowski (Kantola et al. 2005) at the University of Louisville, KY. Cardioid specifies the physical competences of humans.

The database of our resource ontologies should be easily accessible and understandable to all stakeholders and individuals regardless of where they are. The goal is to involve stakeholders in the development and management of the organization and its resources. What this presentation means is described in Section 3.1.5.

3.1.5 Ontology Presentation

We can provide a conceptual basis to support and guide stakeholders, such as managers, in their work with the help of an easily accessible global ontology repository of the resources they utilize. The content and structure of organizational resources is typically not available to people who are doing their work. One problem with ontologies is that the concepts typically are not familiar to people in different work roles. The description and language of ontologies can be quite far from a normal working language in different work roles. Moreover, the presentation of ontologies is typically quite technical, resembling programming or modeling languages that are easily understandable by experts who work with ontologies, but are more difficult to understand for other working people. This means that we need to present organizational resource ontologies in the ontology repository to stakeholders using natural language and such terminology as is familiar to people in a variety of occupational roles and target groups.

Concepts have different kinds of characteristics. These characteristics typically have easily perceivable evidences at workplaces in organizations. These kinds of evidences indicate the presence and status of the concept from a variety of angles in an organization. In this book, these perceivable evidences of the concepts are defined as indicators. Each concept has a number of indicators ranging from a few to many. Some indicators are stronger than others, that is, they provide stronger evidence of the state of the resource; refer to Figure 3.3.

One thing is sure: the indicators are not mutually exclusive, which is a typical prerequirement in mathematics and statistics regarding inputs for the use of certain methods and calculations. In real life, many of these nonmutually exclusive indicators are present at the same time. The indicators need to be specified and presented to stakeholders with such semantics as are familiar to them in their own working context.

TABLE 3.2
Evolute Ontology Repository Contains Different Kinds of Organizational Resource Ontologies

Name	Version	Domain	Current Language Versions
Accord (Tuomainen 2014)	1.0	Company democracy	Eng
Astroid (Taipale 2006)	1.0	Competencies of sales personnel	Fin, Eng, Swe, (Rus), (Cat), (Spa), (Pol), (Kor)
Bicorn (Halima 2007)	1.0	Safety culture	Fin, Eng, Rus, (Pol), (Fre)
Cardioid (Kantola et al. 2005)	1.0	Physical competence	Fin, Eng, Swe, Cat, Spa
Cardioid kids (cf. Kantola et al. 2005)	1.0	Children's physical activity and competences	Fin, Eng, Swe, Cat, Spa
Chronos and Kairos (Reunanen 2013)	1.0	Managing time	Fin, Eng
Cissoid (MSc thesis, Evolutellc.com)	1.0	Maintenance and salesperson combined	Fin, Eng, Swe, (Rus), (Cat), (Spa), (Pol), Kor
Cochleoid (Haaramo 2007)	1.0	Competencies of buyers	Fin, (Swe), (Rus), (Cat), (Spa), (Pol), (Kor)
Conchoid (Mäkiniemi 2004)	1.0	Competencies of maintenance personnel	Fin, Eng, Swe, (Rus), (Cat), (Spa), (Kor)
Cycloid (Liikamaa 2006)	1.0	Competencies of project managers	Fin, Eng, Swe, Rus, Cat, Spa, Pol, (Kor), Fre
Deltoid (Nurminen 2003)	1.0	Competencies of plant operators	Fin, Eng, Swe, Rus, (Cat), (Spa), Pol, (Kor), Rus
Epitrochoid (Lintula 2004)	1.0	Competencies of human resource managers	Fin, Eng, Swe, (Rus), (Cat), (Spa), (Pol), (Kor)
Folium (Paajanen 2006, 2012)	1.0	Knowledge creation	Fin, Eng, (Swe), Cat, Spa, Pol, Kor, (Fre)
Helicoid (Mäkinen 2004)	1.0	Competencies of executives	Fin, Eng, Swe, (Rus), (Cat), Spa, (Kor)
Helix (Einolander and Vanharanta 2014)	3.0	Commitment; Part A: organizational commitment Part B: engagement	Fin, Eng, Pol, Gre, Cat, Spa, (Bas)
Helix Academic (Einolander and Vanharanta 2014)	1.0	Academic commitment	Eng
Kappa (Ingman, Kari—Chapter 6 in this book)	1.0	Sales culture	Fin, Eng
Lame (Jusi 2010)	1.0	Service culture	Fin
Languoid (Odrakiewicz et al. 2009)	1.0	Language studies	Eng
Lissajous (MSc thesis, Evolutellc.com)	1.0	Project business culture	Eng
Lituus (Paajanen 2006; Paajanen et al. 2004)	1.0	Knowledge creation and organizational learning	Eng

(Continued)

TABLE 3.2 (*Continued*)
Evolute Ontology Repository Contains Different Kinds of Organizational Resource Ontologies

Name	Version	Domain	Current Language Versions
Metatrin (Anttila 2013)	1.0	Purchasing	Fin
Nephroid (Hurme-Vuorela 2004)	1.0	Competencies of human work professionals	Fin, Eng, Swe, (Rus), (Cat), Spa, Pol, (Kor)
Pearl (Seikola 2013)	1.0	Customer experience	Eng
Pedal (Vanharanta, thesis, Evolutellc.com)	1.0	Being at work	Fin
Pursoid (Ingman; Vanharanta et al. 2012)	2.0	Innovativeness	Eng
Rhodonea (Aramo-Immonen 2009)	1.0	Megaproject management	Fin, Eng
Rose (Vanharanta et al. 2007)	1.0	Love dimensions	Fin, Eng, Swe
Serpentine (Piirto 2012; Salo 2008)	2.0	Safety culture	Fin, Eng, Spa, Fre, Gre, Cat, Spa, (Bas)
Sinusoid	1.0	Work role selection	Eng
Spiric (Kivelä 2011)	1.0	Growth company—Parts A and B	Fin, Eng, Swe
Spiricoid (MSc thesis, Evolutellc.com)	1.0	Project requirements' engineering	Eng
Strategus (Vanharanta and Kantola 2015)	1.0	Strategy	Eng
Strophoid (MSc thesis, Evolutellc.com)	1.0	Financial controllers' competences	Eng
Talbot (Paajanen 2006, 2012)	1.0	Learning environment	Fin, Eng, Swe, Cat, Spa, Pol, Kor
Tractrix (Naukkarinen et al. 2004)	1.0	R&D project portfolio	Eng, Rus
Tricuspoid (Palonen 2005)	1.0	Entrepreneurs' competences	Fin, Eng, (Rus), (Cat), (Spa), (Pol), (Kor)
Trifolium (Heikkilä 2005)	1.0	Supply and value chain management	Fin, Eng, Swe

Fin, Finnish; Eng, English; Swe, Swedish; Rus, Russian; Cat, Catalan; Spa, Spanish; Pol, Polish; Kor, Korean; Fre, French; Gre, Greek; Bas, Basque; (), partially.

The point is that stakeholders should be able to perceive indicators easily in the course of their normal work, operation, and business.

Indicators presented in natural language form one or more semantic layers between people and concepts, that is, between the conceptual world and the easily perceivable world. According to Gillette, phenomena are those things that appear to be. Phenomena appear as knowledge in perception, as they are perceived, and

TABLE 3.3
Cardioid Describes the Conceptual Structure of Physical Competences of Humans

Physical Competences	Competence Group	Competence Main Groups
Strength, flexibility, endurance, dexterity	Physical ability	Motor competences
Coordination, motor control ability, speed accuracy trade-off, and task performance under stress	Psychomotor ability	
Aerobic capacity	Cardiovascular/respiratory ability	
Motivation to use physical resources, feeling of well-being, physical ability, self-awareness of one's physical abilities, and exercise habits (sports)	Management of physical abilities	
Knowing what to eat, eating well, and controlling body weight	Nutrition	Nutrition competences
Visual, auditory, vertical balance, and thermal regulation	Perceptual system	Environmental awareness competences

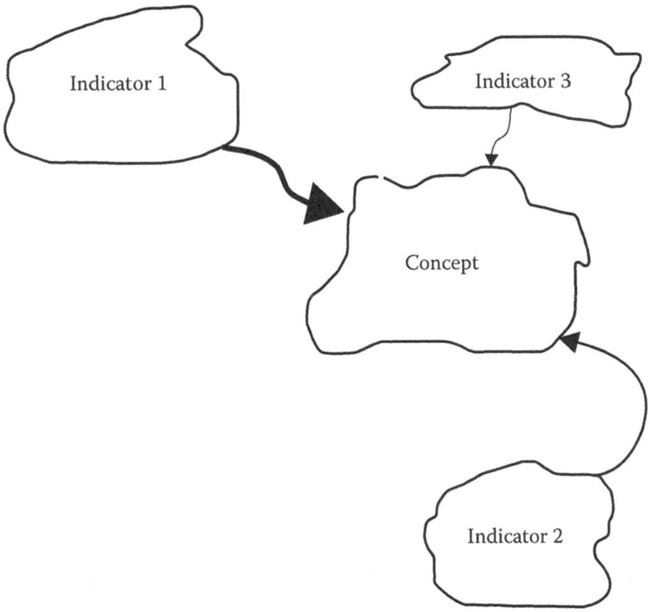

FIGURE 3.3 Indicators are perceivable evidence of concepts in the specific context in an organization.

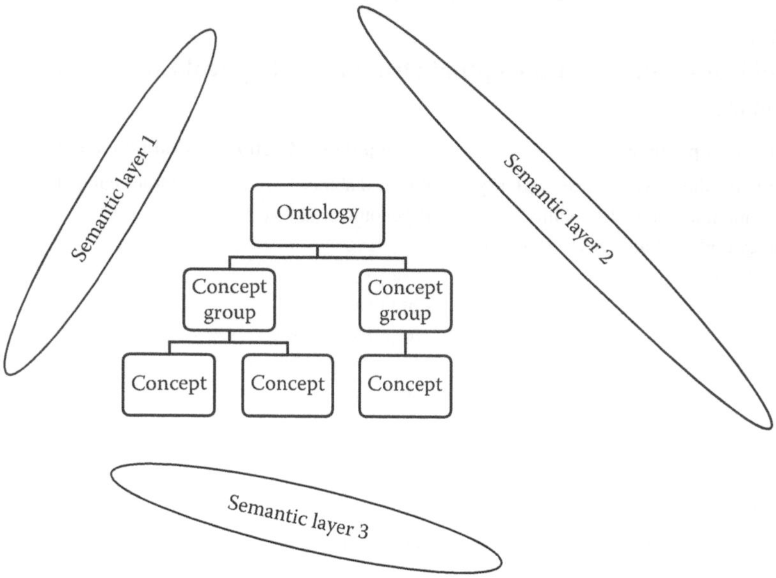

FIGURE 3.4 Semantic presentation of the ontology depends on the target group.

facts are representation of phenomena (Gillette 2000). Facts represent phenomena. Indicators represent concepts. Such semantic layers may be multiple and they may be different and overlapping; refer to Figure 3.4. This means that the organizational resource ontology can be the same, but the sets of indicators presented to different stakeholders may be different. Furthermore, there are even several semantic layers (sets of indicators) of the same ontology. Conceptual comparisons of views on the organizational resources can be done in different organizations. Semantic layers can be in different languages as well. Still, the common ontology denominator is the same, allowing for comparative and longitudinal observation of organizational resources. Therefore, it is important to develop domain ontologies that are as generic as possible instead of developing a unique ontology in each case. Not having a common ontology denominator entails difficulty in comparing different cases.

Presenting the semantic layers to stakeholders can be done with contemporary and mobile ICT solutions. The goal is to involve stakeholders and capture their expertise for the management process of the organization with the help of indicators. After presentation, there is a need to collect and interpret the meaning of people's perceptions in terms of the concepts in the ontology. A collection and interpretation system is needed. It will be described later in this book.

3.1.6 Ontology-Based Self-Evaluation

Self-evaluation is an efficient method to develop oneself, manage personal growth, clarify roles, and commit to project-related goals (Nurminen 2003). Self-evaluation is not accurate (Beardwell and Holden 1995), but it is possible to evaluate oneself with a certain degree of accuracy. The question is whether individuals want to evaluate themselves, and whether they will do it with sufficient care so as to be beneficial. It is acknowledged that self-evaluation is more effective in evaluating the relation between different items (e.g., Torrington and Hall 1991) than evaluating work performance or comparing individuals' performance to that of others (Dessler 2001). This is due to personal bias in seeing one's own performance as better than others'. The effectiveness of self-evaluation depends, for example, on the content of the evaluation, the application method, and the organizational culture (Torrington and Hall 1991). The results of self-evaluation vary to some extent (Cronbach 1990). In the short term, the results change with fluctuations in the individual's power of observation, intentions and motives. In the longer term, the results reflect mental growth, learning, and changes in personality and health.

Self-evaluation also refers to the evaluation of those systems that the evaluator is part of. This is often the case with organizational resources. In this book, ontology-based self-evaluation refers to the evaluation of the indicators of internal and external concepts in organizational resource ontologies. On other words, it refers to the evaluation of the semantic layers of organizational resources (Section 3.1.5). Figures 3.5 and 3.6 illustrate self-evaluation with indicators. The perceived current state and envisioned future state of indicators are evaluated. In Figure 3.5, an internal concept is evaluated, while in Figure 3.6 an external concept is evaluated.

Ontologies are perceived and evaluated in dynamic situations. This offers promise for developing adaptive products and services that behave and manifest themselves differently in different situations for different users.

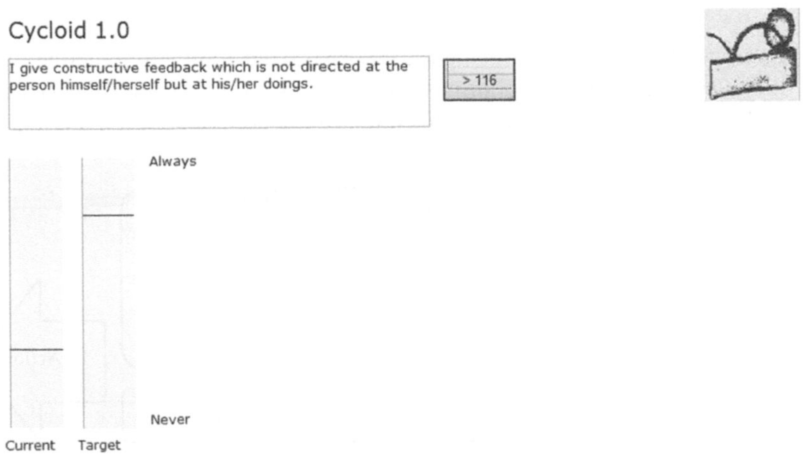

FIGURE 3.5 Evaluating competence indicators of the cycloid tool (Liikamaa 2006) with the Evolute system.

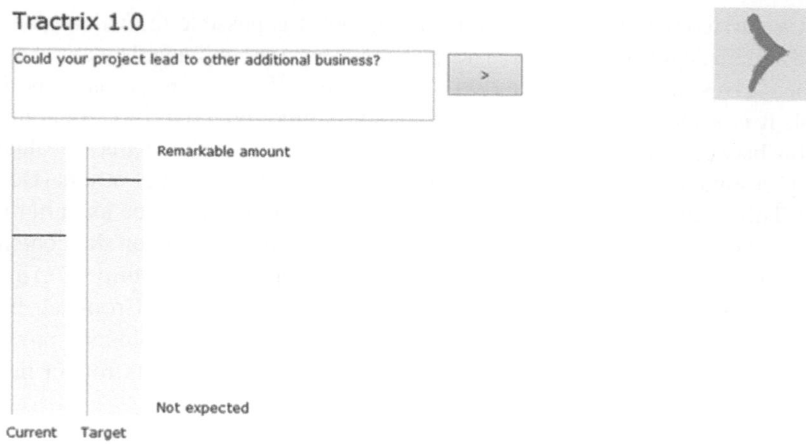

FIGURE 3.6 Evaluating R&D project portfolio indicators of the Tractrix tool (Naukkarinen et al. 2004) with the Evolute system.

3.2 PROPOSITION 2: INVOLVE PEOPLE WITH MODULAR PROCESS

The second proposition says that stakeholders need to be involved in the management of the organization. In the Evolute approach, this is done through a modular process that shows what kind of role stakeholders have in different parts of the management process. Figure 3.7 shows these modules and how people are involved in the management process according to the Evolute approach. Each module has its own important role in the Evolute approach. The role of each module is explained after the figure.

1. *Plenary*: Theories, methodologies, methods, and tools are explained in the organization at the beginning of the process. From our long experience in many different kinds of organizations, we know there are many fears

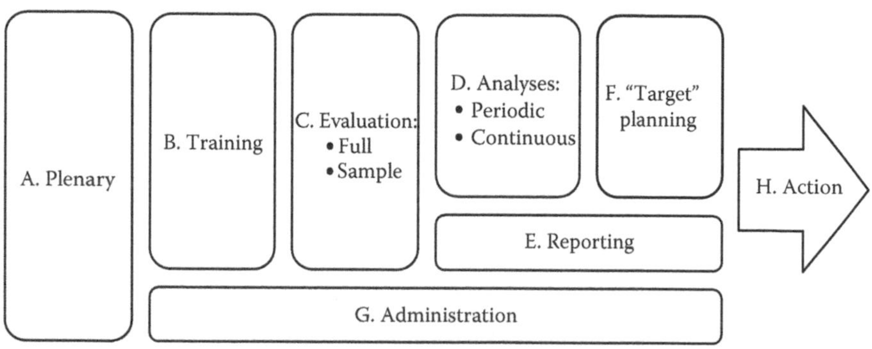

FIGURE 3.7 Modules for involving people according to the Evolute approach.

that people have regarding any new system that is introduced to them. It is crucial to the success of the effort to have plenary (opening) sessions where everything is explained in a positive manner. What is the goal of this kind of effort in the organization? What is self-evaluation? And so on. People have fears about where their data will be used and whether their self-evaluations are going to be compared against those of others. But, based on our experience, when the process of involvement and the role of people as experts to participate in the development or organizational resources is explained, typically these fears have gone away. It is also crucial to emphasize how strict the privacy requirements are. Then, fears have turned into anticipation at being able to participate. It has become very clear that it is difficult to have a successful start with virtual presence only. Plenary information sent in e-format does not make people's fears go away. The personal presence of the management is required at the beginning.

2. *Training*: Just like with any other new system, the user of the system needs to be trained using the "train the user and admin" or "train the trainer" principle. The goal has been to make the tools easy to use, so that there is almost no technical training required. Training addresses the following questions:
   a. How are organizational resource development projects administered so that stakeholders are actively involved and their expertise is analyzed and reported?
   b. How are the results used in the management effort in the organization?
   c. What is the schedule for the modules?
   d. How is the Evolute system used?
3. *Evaluation*: In this module, evaluations (self-evaluations) are conducted online within a given time window by using any device such as a computer, tablet, smartphone, or something else that allows the collection of required inputs. Indicators can be presented in a certain order or in a random order. Everyone in a certain stakeholder group can do the evaluation or a selected sample or random sample of people from a certain stakeholder group can do the evaluation. All these evaluations produce a raw dataset of input values of indicators. There are different ways this can be done:
   a. Full evaluation refers to perceiving and answering all indicators of the used tools.
   b. Sample evaluation refers to perceiving and answering a selected or random set or sample of indicators from the set of all indicators.
4. *Analyses*: This module refers to analyzing accumulated inputs from stakeholders. Analyses produce datasets that are computed outcomes from the Evolute system based on the input dataset and organizational resource ontologies. The analyses can be stored in local databases, servers, or the cloud. Different types of analyses that are required depend on the project goals: visual, graphical, numerical, statistical, text-based, and self-organizing maps (SOMs). Analyses are automatically or semiautomatically computed and generated from the specified datasets for management

purposes. Analyses can be generated in periodic cycles or on a continuous basis depending on the need of the management. This module also contains any consulting work needed to analyze datasets.

5. The reporting module relates to producing reports that are required to support management activity. Reports based on the analyses are created and delivered. Individuals receive their own personal report immediately after the evaluation is finished because the computing is quick. Based on one analysis, many different kinds of reports can be generated: numerical, visual, and also verbal. Reporting can be done using the periodic principle, meaning that in the end of a specific project, the reports are created. This can also be done in cycles so that each cycle produces a set of reports. Another way to do the reporting is on a continuous basis. This means that evaluation, analysis, and reporting are integrated parts of work processes, for example, based on selected or random sample principles. Examples of this kind of random sample approach include the following: (1) every morning employees evaluate a few indicators when they log into their organization or work system and (2) every week plant operators evaluate a random set of indicators to activate their *license to work*. In both of these examples, deviations from the so-called normal reports can be monitored. If everything seems fine, the person can access the work system whether it is an office computing environment, a nuclear power plant control room, or a mass transportation system. Unusual reports can easily be monitored using exploratory data analysis tools, such as SOMs by Kohonen (2001). If there is some clear deviation from normal reports, the state of the work system and the person should be checked to ensure normal and safe operation.
6. *Target planning*: In this module, the reports of analyses are used in management work or development projects. From their storage, the reports are accessible during the project. According to our experience on the *Evolute projects*, stakeholders are typically motivated to be involved in the management effort in this module too. Different kinds of workshops, seminars, and meetings can be organized to interpret the reports together. A good way has been to introduce three elements to be used in the workshops: (1) the reports, (2) strategy of the organization, and (3) other relevant materials. Based on these three elements, the workshop participants can see which concepts of the organizational resources require further attention. Thus, the reports enable organizations to make targeted and heuristic management and development plans regarding the organizational resources they utilize. The planning of the targeted effort typically saves money, time, and human resources, since the development efforts can be put in the right place and in the right order. There is no need to guess where to look.
7. *Administration*: This module refers to maintaining the project data and supporting stakeholders with any technical or other challenges to take part in the system.
8. *Action*: In this module, targeted action plans concerning organizational resources are put into practice.

In summary, the modules describe the process that the Evolute approach follows to involve stakeholders in the management of organizational resources in any organization. After plenary sessions for new people, the modules C, D, E, F, and H are repeated in periodic cycles—as will be described in detail later in Section 3.4.

3.3 PROPOSITION 3: COMPUTE THE MEANING WITH FUZZY LOGIC

The ontological forms of organizational resources as such are incomprehensible to most people, since their presentation is typically mathematics-like. That is of course fine from a computing point of view, but people are not good at dealing with such technical format and presentation. Because of this, the format of ontologies presented to people should be visual and linguistic due to the nature of organizational resources, as described earlier. Another reason to prefer linguistic presentation is in the attempt not to lose knowledge provided by people in conversions between linguistic and numeric presentations. Together, ontological and linguistic presentations of organizational resources are a good combination for computation and humans. In order to show ontologies to people in a linguistic format and find out what they mean, a computing system is needed, which understands the mathematical format of organizational resource ontologies, the linguistic format of indicators of these ontologies, and the concepts of these ontologies.

People's ability to make precise yet significant statements about a system's behavior diminishes as the complexity of the system increases (Zadeh 1973). Organizational resources are complex systems and thus precise mathematics does not really work well in this context. Instead, soft-computing is a more suitable approach (Zadeh 2005). The best-known way to reason with linguistic information based on logical constructs today is fuzzy logic (Kosko 1994; Zadeh 1965, 1973). Fuzzy logic has been around a long time (Berkan and Trubatch 1997) and is still one of the fastest growing technologies in the world since the beginning of the computer era. At the end of 2014, there were more than 360,000 publications with "fuzzy" in the title and more than 560,000 fuzzy logic–related patents issued (Zadeh 2014).

This chapter describes how the ontology-based meaning according to the Evolute approach is computed with fuzzy logic. After the stakeholders are involved, the meaning of their knowledge input need to computed. As we remember from earlier sections, the conceptual domain is different from the working domain where the indicators can be perceived. According to the Evolute approach, the meaning is computed with the help of fuzzy sets (Zadeh 1965), linguistic variables (Zadeh 1975), and fuzzy logic (Zadeh 1973). How this is done is explained in the next sections.

3.3.1 Fuzzy Sets

Vagueness in linguistics can be captured mathematically by applying fuzzy sets (Lin and Lee 1996; Zadeh 1965). This is done by creating linguistic variables that *contain* fuzzy sets. Fuzzy sets represent systems better than crisp sets do for two reasons. First, the predicates in propositions representing a system do not have crisp denotations. Second, explicit and implicit quantifiers are fuzzy (Zadeh 1983). A fuzzy set can be defined mathematically by assigning to each possible individual in the universe of discourse a value representing its grade of membership in the fuzzy set. This grade corresponds to the degree to which that individual is similar to or compatible with the concept represented by the fuzzy set (Klir and Yuan 1995). In the Evolute approach, the perception of different indicators becomes a degree of membership in fuzzy sets. Just like things are in real life, everything is a matter of degree regarding organizational resources. Linguistic variables bridge the gap between the mathematical base variable in the universe of discourse and the linguistic meaning in the human mind.

3.3.2 Linguistic Variables

In the Evolute approach, both indicators and concepts in organizational resource ontologies are defined as linguistic variables. Fuzzy sets can capture and model the nature of indicators in a mathematical form that is the base for computations of meaning. As a researcher or practitioner specifies the indicators of a concept in organizational resource ontology, he or she can capture the nature of the indicator in real life using fuzzy sets. Indicators are all different and since they describe real aspects of real objects their mathematical presentation cannot really be standard at all times. The same applies to concepts in organizational resource ontologies as well. In the beginning, trapezoidal and triangular fuzzy sets (membership functions) can be used unless there is better understanding of the nature of the indicator at that time. The Evolute system uses triangular and trapezoidal membership functions in indicators and concepts, but of course S-shaped, bell-shaped, or any other shaped membership functions can be used as well.

Every measure and variable of any metrics that is followed in any organization has the following value ranges: desired range (good), acceptable range (medium), and unacceptable range (low). Figure 3.8 describes two samples of three fuzzy sets in a linguistic variable (indicator). When there is better knowledge and deeper understanding of the nature of the indicator and how it behaves in real life, the good, medium, and low ranges (fuzzy sets) are adjusted accordingly.

The value of a measure in the good range indicates that the target of the measure is in good shape. A value in a low range, in turn, requires immediate attention and

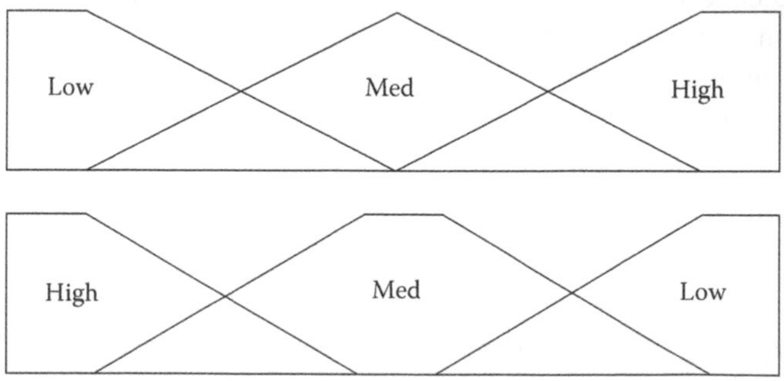

FIGURE 3.8 *Standard* fuzzy sets can model indicators at the beginning of the research.

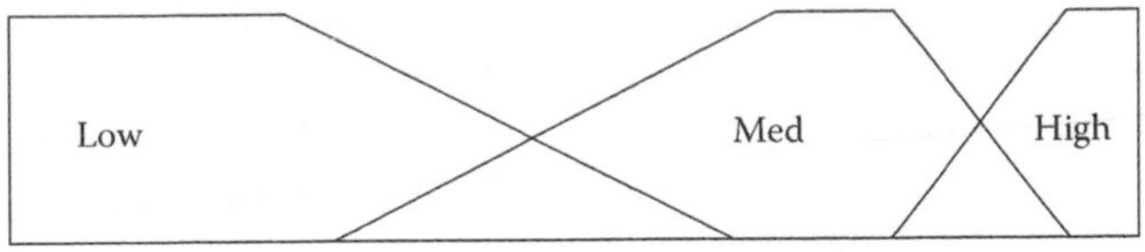

FIGURE 3.9 This indicator has a narrow range of high values and wide ranges of low and medium values.

TABLE 3.4
Indicator Samples from the Ontology Repository

Indicator	Min	Max
Linguistic variables	*Linguistic values*	
Staff initiative for solving safety glitches is	Passive	Active
I act in accordance with my own principles and values even though they are in conflict with the principles and values of the organization.	Always	Never
I hold on to my principles even if they are unpopular.	Never	Always
The access to important job-related information for a new employee is	Difficult	Easy
In my new job the information related to employment (benefits, responsibilities, etc.) became available	Slowly	Fast
Is this project a technology push?	Not at all	Highly
Our local partners in cooperation are	Unreliable	Reliable
My organization learns from errors.	Not at all	Fast
I feel_______in my job.	Incompetent	Competent

action by the management. The acceptable range is somewhere in between; attention is needed but not urgently. Figure 3.9 shows an example of an indicator that has quite a large range of values that are low and medium and quite a narrow range of high values based on the understanding of the indicator in real life.

It is also possible to have the low, medium, and high membership functions in a different order, for example, the high value could be in the middle of the universe of discourse. The membership functions of an indicator really should reflect the observed reality of the indicator. Table 3.4 shows samples of different kinds of indicators. The idea is to have the dimension that has value on the linguistic scale and present it with linguistic labels that show the minimum possible value and the maximum possible value of the variable. The linguistic labels represent the whole possible range of linguistic values the variable can have.

Based on the experience of the Evolute approach, there are several things to pay attention to when specifying indicators as linguistic variables. They are at least the following:

- Choose indicators to describe different perceivable viewpoints on the concept.
- Move the dimension that has changing value to the linguistic scale: level of agreement, frequency, amount, level of truth, any adjectives, etc.

- Make indicators simple and clear. Do not join two linguistic variables in one indicator by using AND/OR operators.
- Avoid indicators that describe the same aspect of the concept and are quite similar.
- Avoid ubiquitous indicators.
- Fewer than 10 indicators should be enough to indicate the state of the concept.
- You can use indicators that have reversed meaning in the linguistic variable or its values. However, these have been observed to be the most difficult to understand according to participants and have created the most confusion and misunderstanding.
- Observe the relative weight of the indicators within concepts based on practical experience and explicit material.
- Observe high, medium, and low ranges for indicators based on practical experience and explicit material.
- Increasing the number of indicators enables the capturing of more tacit knowledge/meaning but increases the time and effort required from the participants who are involved. Decreasing the number of indicators enables capturing less tacit knowledge/meaning but decreases the time and effort required from the participants who are involved.
- Iterate several times.
- The set of indicators should follow the accumulated experience of the domain and changes in a dynamic working environment.
- The opposite of an apple is not an orange or a banana—it is "not apple" (1-apple).

3.3.3 Fuzzy Logic

The Evolute system computes the meaning of how stakeholders perceive the indicators. Fuzzy logic is a precise logic of imprecise things (Zadeh 1973). Fuzzy logic allows reasoning using fuzzy sets and fuzzy rules. It has two principal components (1) a translation system for representing the meaning of propositions and other semantic entities and (2) an inference system to arrive at an answer to a question that relates to the information resident in a knowledge base (Zadeh 1983). Here, propositions refer to the semantic layers (indicators) of the concepts of organizational resources. The knowledge base refers to the concepts (ontologies) of organizational resources. Fuzzy logic can be used to resemble an expert's task of evaluating and reasoning based on linguistic information in applications such as the Evolute system. The Evolute system is this kind of fuzzy logic application. With the help of fuzzy logic, the reasoning ability of domain experts can be captured and embedded in the Evolute system and its ontology repository.

Conventional mathematical methods based on 0–1 logic require that certain preconditions are met before they can be utilized, for example, the concern with the independence of the factors used in many statistical methods. Fuzzy logic allows us to ignore these preconditions due to the use of linguistic variables (Wilhelm and Parsaei 1991). Conventional mathematical methods and statistics encounter difficulties when applied to human beings and human systems with imprecision and uncertainty. The world is not precise enough for many mathematical methods based on Aristotle's 0–1/either–or foundation. As mentioned earlier, everything in real life is a matter of degree.

3.3.4 Mapping Domains

Now, we know that we need to link the perceivable working environment in the organization where the indicators reside with a conceptual domain that describes the holistic content of the organizational resources. According to the Evolute approach, these two domains are mapped with the help of fuzzy logic; refer to Figure 3.10.

The logic in fuzzy logic is normal logic, that is, the logic itself is not fuzzy. Fuzzy refers to inputs and outputs, that is, to the sets that unsharp boundaries (Zadeh 1971). The logic refers to how inputs are related to outputs (Zadeh 1973), that is, which inputs are involved in producing the output or consequence. Fuzzy logic determines which antecedents are mapped with which consequents. Antecedents are indicators and consequents are the concepts in ontologies (top-level concepts, concept groups, and concepts). Figures 3.11 through 3.14 illustrate different options of this mapping in the Evolute system. These options form the basis of the logical part in fuzzy logic.

The observation of indicators and their strength in real life is important, since some indicators are stronger than others in indicating the state of the organizational

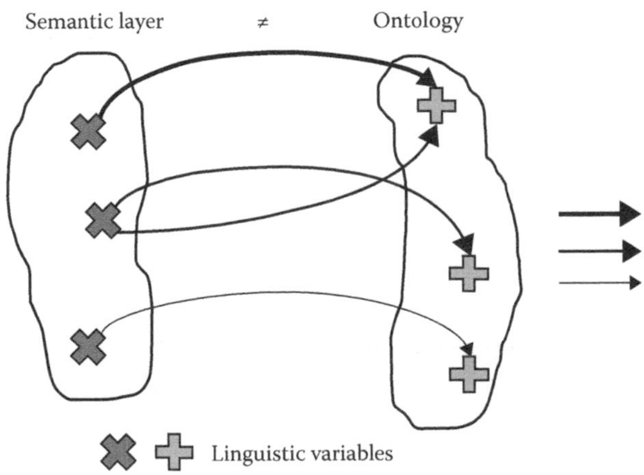

FIGURE 3.10 Two different domains are mapped using fuzzy logic.

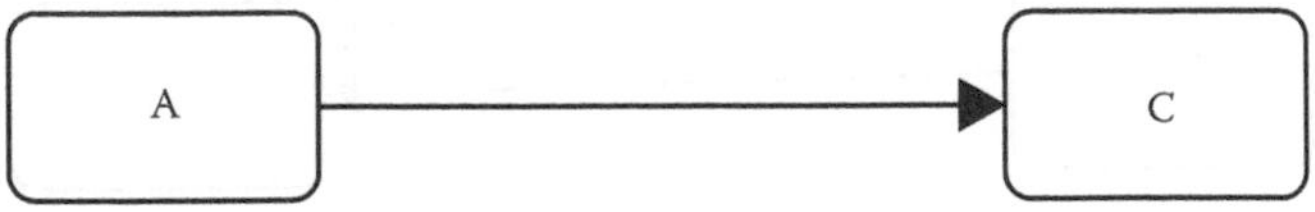

FIGURE 3.11 One antecedent (A) to one consequent (C) mapping.

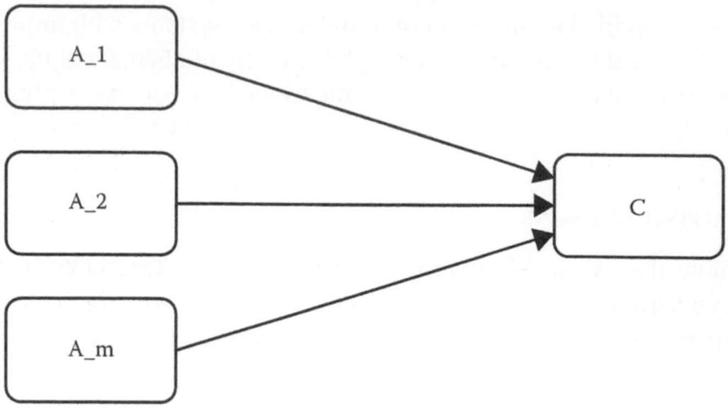

FIGURE 3.12 Many antecedents to one consequent mapping.

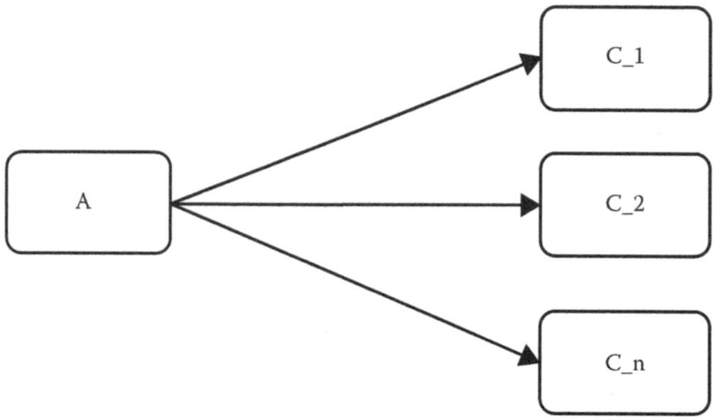

FIGURE 3.13 One antecedent to many consequents mapping.

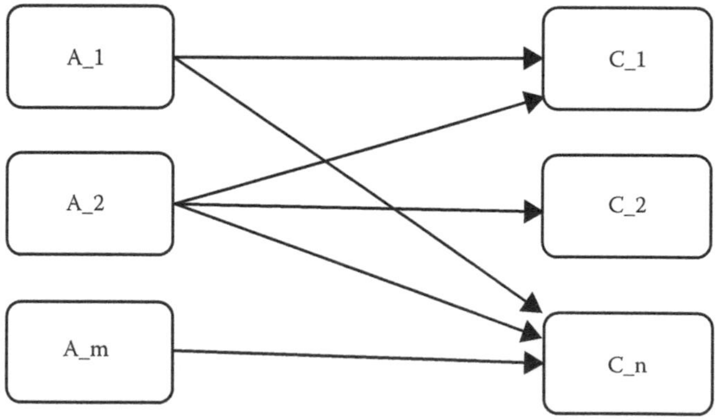

FIGURE 3.14 Many antecedents to many consequents mapping.

resource. It is possible to specify which indicators are stronger than others based on observation, research, and experience. It is also possible to specify what kind of linear or nonlinear logic the indicators of the concept have together. Each concept and its indicators form truly a unique whole. This means that the indicators of the concept and their relative weights only apply within each concept alone. For example, indicator X indicates the state of two concepts A and B so that X is the strongest indicator for A but the weakest for B. The relative strength (weight) of indicators is illustrated in Figures 3.15 through 3.19.

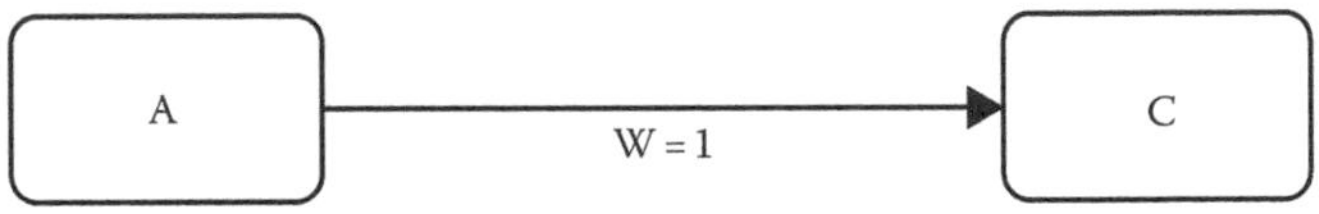

FIGURE 3.15 One antecedent with a weight of 1.

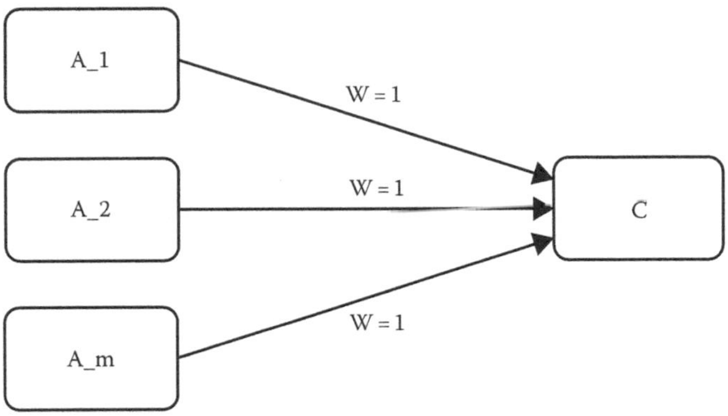

FIGURE 3.16 Many antecedents with equal relative weight within the concept.

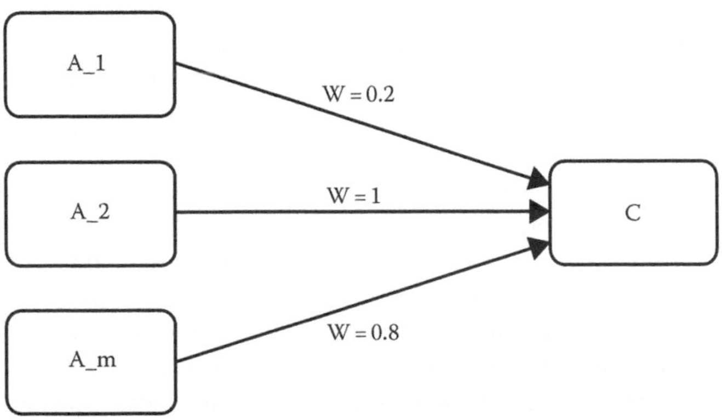

FIGURE 3.17 Many antecedents with different relative weights within the concept.

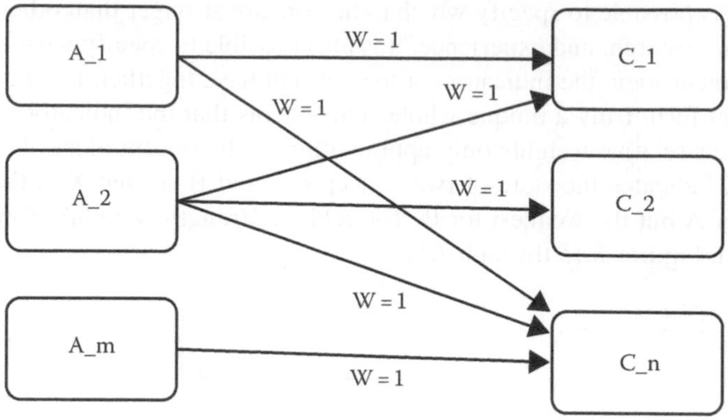

FIGURE 3.18 Many antecedents mapped with many consequents with equal relative weights.

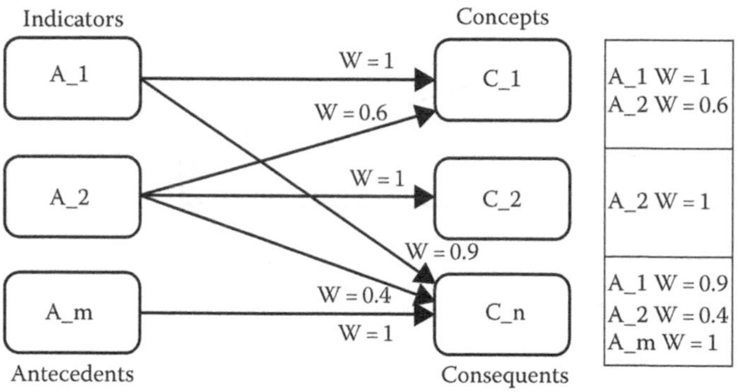

FIGURE 3.19 Many indicators mapped with many concepts with different relative weights for each concept.

The different kinds of mappings shown in Figures 3.11 through 3.19 between the indicators and the concepts of organizational resources are based on the observations of the reality by the researcher/practitioner. These kinds of observations of causal connections are always necessary regardless of the research type: new research or re-research. Typically, there is not much explicit material available about the semantics and ontologies of organizational resources that are being studied. In terms of research, this can be exploratory research (Saunders et al. 2007), and reasoning is inductive or *bottom-up* when tools, such as Evolute tools, are being developed.

When the mapping of antecedents and consequents is underway in new organizational resources research, it can be difficult to assign relative weights for antecedents within a concept. If the researcher is an expert on the domain, he or she can understand the causality so well that he or she is able to assign relative weights to the antecedents. The more research and practical expertise there is available of the domain, the more realistic the mapping of indicators and concepts with relative weights can be.

This follows the normal development process of ontologies where empirical testing and adjustment are iterative parts of the development process. This includes antecedents, consequents, and their mapping. In other words, it includes fuzzy sets and fuzzy logic. This means that over time, when more experience of organizational resources is accumulated, more reliable models can be made. Understanding the mapping of indicators and concepts of organizational resources, as described in this chapter, serves as a base to develop fuzzy rule bases to specify the causality between the semantic indicator domain and conceptual ontology domain. How this kind of fuzzy rule base can be created is explained in Chapter 4. One can ask, what is the point of trying to capture the reasoning capability of a domain expert with the fuzzy logic that is applied in the Evolute system? There are important reasons for doing this: experts are not available everywhere anytime when needed. Most people at work are not domain experts. Moreover, the conceptual reality is much more complex than is typically understood and therefore not even an expert always has a holistic base for reasoning regarding the organizational resource in focus. Lastly, the outcome of the reasoning is immediately available.

3.3.5 Rule Bases

A fuzzy rule base is a collection of IF–THEN rules in which the preconditions and consequences involve linguistic variables (Lin and Lee 1996). In this context, the rule base characterizes the input–output relation of the individual concept of an organizational resource, that is, the indicator–concept relationship of an organizational resource. The following general design guidelines apply to the rule bases (Silva 1995):

- The rule base should be complete. All possible input combinations should be covered and the inputs should adequately represent the problem. One problem with fuzzy logic controllers (FLCs) is to ensure the *completeness* of the performance of the FLC (Sayyarrodsari and Homaitar 1997).
- A rule base should be continuous. When the condition's fuzzy sets overlap, the action fuzzy sets also overlap.
- A rule base should be consistent. There are no contradictory rules; that is, for the same antecedent, two consequents are possible.
- The rules should not *interact*. If the rules interact, the antecedent (premise) of a particular rule, when composed by the rule base, may not necessarily result in the consequent (conclusion) as given by the rule itself (Pedrycz 1989).

In the fuzzy rule bases used in the Evolute system, all possible input combinations are covered and valuable feedback about the *completeness* of current concepts of the organizational resource is collected in user tests and real-world cases. The rule bases used in the Evolute system use the AND operator to cover all possible input combinations. Each concept of an organizational resource has its own rule base in the ontology, which may have anything from a few to millions of rules. Rule bases, the number of inputs for each rule base, and the number of decision rules in the cardioid tool are shown in Table 3.5. The number of the decision rules in a rule base is

TABLE 3.5
Number of Decision Rules in the Cardioid Rule Base in Evolute System

Concept/Rule Base in the Cardioid Ontology	Number of Inputs (Indicators)	Number of Rules
Motor competences/physical ability/strength	6	729
Motor competences/physical ability/flexibility	6	729
Motor competences/physical ability/strength/endurance	5	243
Motor competences/physical ability/strength/dexterity	2	9
Motor competences/psychomotor ability/coordination	2	9
Motor competences/psychomotor ability/motor control ability	4	81
Motor competences/psychomotor ability/speed accuracy trade-off	4	81
Motor competences/psychomotor ability/task performance under stress	2	9
Motor competences/cardiovascular–respiratory ability/aerobic capacity	5	243
Motor competences/management of physical abilities/motivation to use physical resources	7	2,187
Motor competences/management of physical abilities/feeling of well-being	10	59,059
Motor competences/management of physical abilities/physical ability	3	27
Motor competences/management of physical abilities/self-awareness of one's physical abilities	8	6,561
Motor competences/management of physical abilities/exercise habits (sports)	3	27
Nutrition competences/nutrition/knowing what to eat	3	27
Nutrition competences/nutrition/eating well	4	81
Nutrition competences/nutrition/controlling body weight	4	81
Environmental awareness competences/perceptual/visual	4	81
Environmental awareness competences/perceptual/auditory	5	243
Environmental awareness competences/perceptual/vertical balance	1	3
Environmental awareness competences/perceptual/thermal regulation	5	243

defined in the following way: (number of linguistic states) ^ (number of inputs), that is, 3 ^ (number of inputs) according to the Evolute system. Not all of the rules fire, of course.

Table 3.5 shows that in the cardioid rule bases the concepts have from 1 to 10 inputs. These inputs are indicators that are linguistic variables. The combinations of these antecedents cover all possible input combinations. A typical way to set and present simple rule bases is in a table format. Table 3.6 shows a simple rule base with two antecedents. The values of inputs can be low, medium, and

TABLE 3.6
Rule Base with Two Antecedents Describes One Concept in Cardioid

	Antecedent 2		
Antecedent 1	**Low**	**Med**	**High**
Low	Low	LtM	Med
Med	LtM	Med	LtH
High	Med	LtH	High

TABLE 3.7
A Quantified Rule Base

	Antecedent 2		
Antecedent 1	**1**	**2**	**3**
1	1	1.5	2
2	1.5	2	2.5
3	2	2.5	3

high and the values of outputs can be low, lower than medium (LtM), medium, lower than high (LtH), and high. This kind of rule base where the consequent is the linguistic *average* of the antecedent values can be typically set in the early parts of the research process when there is no deep understanding of the object to create specific rules. According to Kosko (1994), the fuzzy patch is large in the beginning. Another way to present the rule base in Table 3.6 is to quantify the antecedents so that low = 1, med = 2, and high = 3. Then, the rule base looks like it does in Table 3.7.

In the aforementioned two rule bases, the antecedent values are *averaged* and the result value in the rule base is the consequent value. This is typically done when there is no deeper understanding or data about the nature and behavior of the object. As the research proceeds, more specific rules can be created. Additionally, operators other than AND can be used, like OR and NOT. Rule bases presented in numerical format, as in Table 3.7, can be guided and managed on the so-called metalevel. This means that metalevel rules can be created to guide the behavior of the rule base. There are different cases where such metarules can be applied:

- Certain antecedents have more weight than others in the rule base—especially when the fuzzy patch is still large.
- There is a need to make the rule base emphasize certain antecedents in certain contexts. This refers to the context-dependent rule base and behavior. Depending on what the operational context is, the rule base can be tuned

and guided. Products and services can be made to operate differently in different contexts, for example, remotely.

- The rule base has a certain type of general style, that is, aggressive (high consequent values are reached easily) or sluggish (it is more difficult to reach high consequent values).

In the Evolute system, a starting point is that a researcher begins to explore an object domain that has not been explicitly specified earlier as an ontology. Organizational resources, conceptually understood, are the kinds of unexplored objects that we have discussed earlier in this book. There are also no data available that could be used to train rule bases between antecedents and consequents, for example, with the help of neural networks. One more aspect to this is that as a researcher studies the object more, he or she will gain better understanding about the behavior of the object. At some point, there is enough experience and data that a rule base with fewer and more specific rules can be specified. From the author's experience, this can take years of research. In technical domains, this takes much less time due to access to measured input–output data. In human systems domains, this takes a longer time due to the lack of measured input–output data that can be used for rule base training and learning. In the Evolute system, this learning curve has been assisted with the help of the metarules that were mentioned earlier. After some time studying the new object, a researcher may be able to say which antecedents are stronger indicators of the nature and behavior of the object, in other words, which antecedents have more relative weight than others within the concept. In the Evolute system, this learning experience is transferred to the rule bases using metarules. Creating fuzzy rule bases is always a matter of teamwork between someone who knows the organizational resource domain and someone who knows fuzzy logic.

Compensation coefficient α can be determined as a factor incorporating the compensatory effect of inputs (Kantola 1998). Alpha (α) is the *average* of quantified antecedents. Alpha (α) can be converted into consequents as shown in Table 3.8. The table shows three different styles of behavior of the rule base: traditional, sluggish, and aggressive. These styles are samples and many other styles can be thought of. With the help of alpha, different kinds of metarules that guide the behavior of the

TABLE 3.8
Sample Determination of the Consequents Based on the Compensation Coefficient α

Consequent	Traditional α	Sluggish α	Aggressive α
Low	$\alpha \le 1.4$	$\alpha \le 1.6$	$\alpha \le 1.2$
LtM	$1.4 < \alpha \le 1.8$	$1.6 < \alpha \le 2.0$	$1.2 < \alpha \le 1.6$
Med	$1.8 < \alpha \le 2.2$	$2.0 < \alpha \le 2.4$	$1.6 < \alpha \le 2.0$
LtH	$2.2 < \alpha \le 2.6$	$2.4 < \alpha \le 2.8$	$2.0 < \alpha \le 2.4$
High	$2.6 < \alpha$	$2.8 < \alpha$	$2.4 < \alpha$

rule base can be applied and lower and upper boundaries for consequents can be set numerically.

Tables 3.9 and 3.10 show a sample rule base with two antecedents: cardioid and dexterity. The first antecedent has a relative weight of one, and the second antecedent has a relative weight of two. The first step is to calculate the consequent (Table 3.9), and the second step is to normalize the result (Table 3.10) back to the original range of 1–3. Table 3.11 shows how the consequents are converted back to linguistic values.

TABLE 3.9
Computing Consequents Using Alpha and the Relative Weights of Antecedents

Cardioid/Dexterity	Antecedent 2 (Weight 2)		
Antecedent 1 (Weight 1)	1(2)	2(2)	3(2)
1	1.5	2.5	3.5
2	2	3	4
3	2.5	3.5	4.5

TABLE 3.10
Normalized Consequents

Cardioid/Dexterity	Antecedent 2 (Weight 2)		
Antecedent 1 (Weight 1)	1	2	3
1	1	1.67	2.33
2	1.33	2	2.67
3	1.67	2.33	3

TABLE 3.11
Computed Consequents are Converted Back to Linguistic Labels According to "α Traditional"

Cardioid/Dexterity	Antecedent 2 (Weight 2)		
Antecedent 1 (Weight 1)	Low	Med	High
Low	Low 1	LtM 1.67	LtH 2.33
Med	Low 1.33	Med 2	High 2.67
High	LtM 1.67	LtH 2.33	High 3

Indicator	Relative Weight	Liguistic Values of Indicators and Concepts																										
A	1	L	L	L	L	L	L	L	L	L	M	M	M	M	M	M	M	M	M	H	H	H	H	H	H	H	H	H
B	1	L	L	L	M	M	M	H	H	H	L	L	L	M	M	M	H	H	H	L	L	L	M	M	M	H	H	H
C	1	L	M	H	L	M	H	L	M	H	L	M	H	L	M	H	L	M	H	L	M	H	L	M	H	L	M	H
A	1.5	1.5	1.5	1.5	1.5	1.5	1.5	1.5	1.5	1.5	3	3	3	3	3	3	3	3	3	4.5	4.5	4.5	4.5	4.5	4.5	4.5	4.5	4.5
B	1	1	1	1	2	2	2	3	3	3	1	1	1	2	2	2	3	3	3	1	1	1	2	2	2	3	3	3
C	4	4	8	12	4	8	12	4	8	12	4	8	12	4	8	12	4	8	12	4	8	12	4	8	12	4	8	12
	Sum	6.5	10.5	14.5	7.5	11.5	15.5	8.5	12.5	16.5	8	12	16	9	13	17	10	14	18	9.5	13.5	17.5	10.5	14.5	18.5	11.5	15.5	19.5
	α = sum/(W1 +W2 + W3)	1	1.6	2.2	1.2	1.8	2.4	1.3	1.9	2.5	1.2	1.8	2.5	1.4	2	2.6	1.5	2.2	2.8	1.5	2.1	2.7	1.6	2.2	2.8	1.8	2.4	3
Rb style	Sluggish	L	L	M	L	LtM	M	L	LtM	LtH	L	LtM	LTH	L	LTM	LtH	L	M	LtH	L	M	LtH	L	M	LtH	LtM	M	H
	aggr.	L	LtM	LtH	L	M	LtH	LtM	M	H	L	M	H	LtM	M	H	LtM	LtH	H	LtM	LtH	H	LtM	LtH	H	M	H	H

L=LOW | LTM=Lower Than Medium | M=MEDIUM | LTM=Lower Than High | H=HIGH

FIGURE 3.20 Sample of computing consequent using alpha in the rule base with three indicators, for example, Cardioid/Exercise habits (sports).

Now when we apply the correlation coefficient alpha, we can see how the formation of consequent is no longer a symmetrical process. We can see that the heavier antecedent is more powerful in determining the consequent than the other antecedent. This means that in real life when applied the input values for the heavier indicator provide stronger causality to the concept of the organizational resource. The same approach can be applied for any rule base with many more than two inputs. Figure 3.20 illustrates the approach with a rule base that has three indicators.

In the Evolute system, the maximum number of antecedents is currently 15. Based on different antecedent value combinations in the rule bases, the linguistic consequents of the decision rules for all rule bases are determined. This means that the rule bases are not hardcoded in the database but instead are created dynamically when needed. When the researcher has enough experience of the objects, he or she can define fewer rules that are more specific. This reduces the amount of time required for computing. Section 3.3.6 describes how rules are used to come up with the meaning of organizational resources.

3.3.6 Fuzzy Logic Application Modules

In the Evolute system, the structure of a general FLC was applied. In general, the more an FLC resembles the expert's role in a (control) task, the higher the implementation benefit will be, that is, in this context, the fuzzy application in the Evolute system resembles managers' task of evaluating the state or *quality* of organizational resources (Kantola 1998; Sayyarrodsari and Homaitar 1997). A general FLC consists of four modules (Klir and Yuan 1995; Mamdani and Assilian 1975) (1) defuzzification module, (2) fuzzy inference engine, (3) fuzzy rule base, and (4) defuzzification module. The interconnection between these modules in the Evolute system is shown in Figure 3.21.

The Evolute system utilizes fuzzy logic to capture the subjective, abstract, and vague nature of the organizational resource ontologies without the individual having to convert any of this on a numerical scale. The goal is to capture a true bottom-up

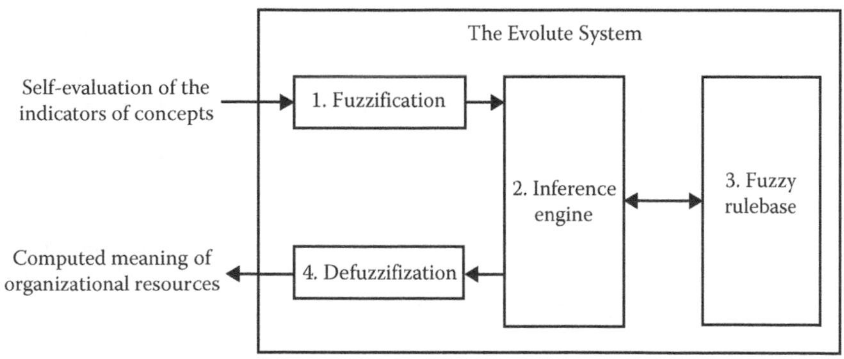

FIGURE 3.21 Interconnection between modules in the fuzzy application applied to the Evolute system. (From Kantola, J., Ingenious management, Doctoral thesis, Pori, Finland, Tampere University of Technology, 2005.)

view of the current reality and envisioned future of the features and practices of organizational resources. The Evolute system works as a generic fuzzy rule base system in the following way (Kantola 1998):

Fuzzification: The evaluation of indicators describing the features of the ontology. Inputs from the evaluation are converted into fuzzy sets. Figure 3.22 illustrates the fuzzification method used in Evolute system (cf. Kantola 1998).

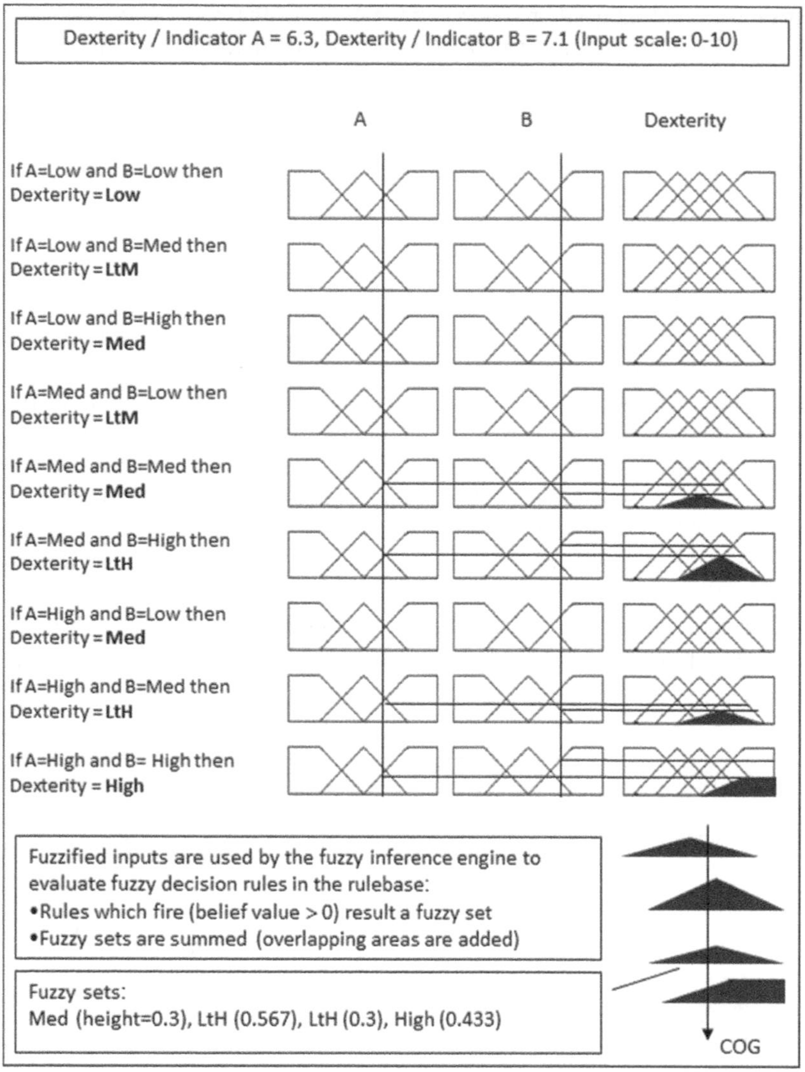

FIGURE 3.22 Illustration of Max-Dot inference method. (Modified from Kantola, J., *A Fuzzy Logic Based Tool for the Evaluation of Computer Integrated Manufacturing, Organization and People System Design*, University of Louisville, Louisville, KY, 1998, 183pp.)

Inferencing: Fuzzified inputs are used by an inference engine to evaluate fuzzy decision rules in the fuzzy rule base. Some rules fire and some do not. This results in one fuzzy set for each concept in the ontology. Figure 3.22 illustrates the Max-Dot inference method used in Evolute system (cf. Kantola 1998).

Defuzzification: Fuzzy sets are aggregated and converted into crisp values that represent the meaning of the perception of the organizational resource by the individual, that is, the state of the organizational resource. The center of gravity (COG)/center of area (COA) defuzzification method that is used in the Evolute system is illustrated in Figure 3.23.

Defuzzified meaning is presented visually and numerically for decision making. As soon as the intended set (full or sample) of indicators is evaluated, the meaning can be immediately computed. Typically, this takes only a few moments, but with larger ontologies the computing may take as long as a minute at present on a standard server. Of course, this depends on the processing power of the server. In summary, with the help of fuzzy logic the meaning can be computed instantly. Section 3.4 introduces instances and explains how the computed meaning is used for management purposes.

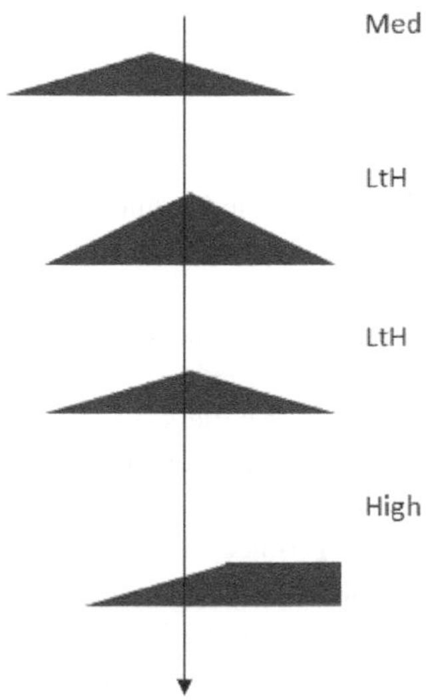

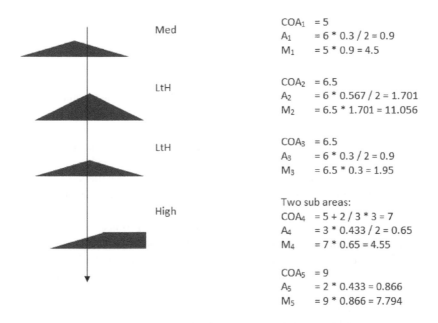

$COA_1 = 5$
$A_1 = 6 * 0.3 / 2 = 0.9$
$M_1 = 5 * 0.9 = 4.5$

$COA_2 = 6.5$
$A_2 = 6 * 0.567 / 2 = 1.701$
$M_2 = 6.5 * 1.701 = 11.056$

$COA_3 = 6.5$
$A_3 = 6 * 0.3 / 2 = 0.9$
$M_3 = 6.5 * 0.3 = 1.95$

Two sub areas:
$COA_4 = 5 + 2 / 3 * 3 = 7$
$A_4 = 3 * 0.433 / 2 = 0.65$
$M_4 = 7 * 0.65 = 4.55$

$COA_5 = 9$
$A_5 = 2 * 0.433 = 0.866$
$M_5 = 9 * 0.866 = 7.794$

COG / COA = $(M_1 + M_2 + M_3 + M_4 + M_5) / (A_1 + A_2 + A_3 + A_4 + A_4) =$

$(4.5+11.056+1.95+4.55+7.794) / (0.9+1.701+0.9+0.65+0.866) = 5.950$

FIGURE 3.23 Illustration of the centroid defuzzification method. (Modified from Kantola, J., *A Fuzzy Logic Based Tool for the Evaluation of Computer Integrated Manufacturing, Organization and People System Design*, University of Louisville, Louisville, KY, 1998, 183pp.)

3.4 PROPOSITION 4: MANAGE IN THE CONTEXT WITH INSTANCES

The fourth proposition says that management must take stakeholders' perception of organizational resources into account in order to manage the organization successfully. With the help of the Evolute system, people can be globally involved and provide their perception of organizational resources and its computed meaning for management purposes. Perception is a holistic composition of several aspects such as one's knowledge and experience of the organizational resource; the big picture and the overall situation; specific organizational context of the resource; personal context; ontology model of the organizational resource; personal values, attitudes, and aspirations regarding the organizational resource; history; and current reality and anticipated future of the resource. All these aspects are embedded in the perception of the resource by a person that is captured in a specific situation, and further computed for management use. In this book, the combination of all these aspects is embedded in and defined as an instance. Instances are explained in Section 3.4.1.

3.4.1 Instances

It is very important that organizations involve their people and learn to utilize their knowledge assets. Several authors write about this, for example:

- The values of those affected by the decision should be taken into account in a decision-making process (Ackoff 1986).
- Firms are differentiated by their ability to leverage knowledge and intellect in creating greater creative value than by an exclusive focus on exploiting physical assets and skills (Rastogi 1995).
- Companies will survive only if they meet the needs of the individuals who serve in them: people's true inner needs (Harvey-Jones 1986).
- Organizations are made up of people whose values and beliefs inescapably influence their thought and actions (Davenport and Prusak 2000).

Every individual has a different basis for understanding objects. This is because the neurons and their connections are constantly shaped by the outside world and therefore everyone's brain is different (O'Connor and McDermott 1997). What people already have in their heads determines how they assimilate new experiences (Leonard and Swap 2004). Without frameworks, domain knowledge, or prior experiences, people may not be able to perceive and process information. Since these receptors are based on an individual's accumulated experience in life, we can say that each individual has a different basis to perceive and understand objects, such as organizational resources. Thus, we can see that no other person can know or communicate how a person perceives an object—an organizational resource in this case.

Within an organization people constantly perceive organizational resources, such as their coworkers. People have personal needs and needs regarding these resources, which are based on their perception of them. Therefore, perceptions play a very important role in an organization's management since people think and act

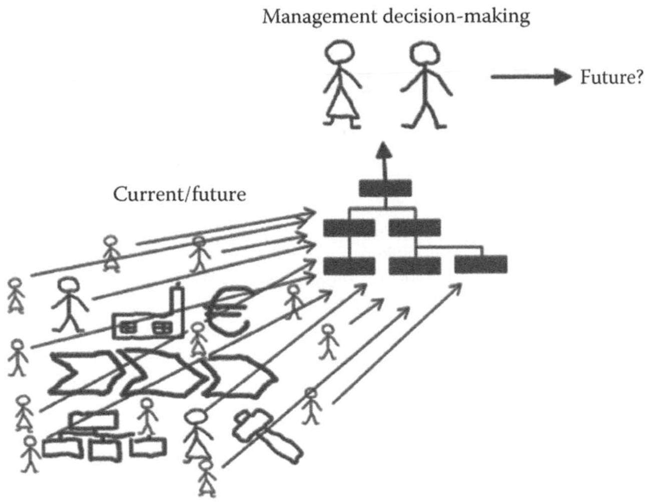

FIGURE 3.24 Stakeholders perceive organizational resources through ontology models for management decision-making.

based upon their perception. Perceptions turn into thinking and finally into action (1) think, (2) talk, and (3) act. People can also envisage the future of organizational resources—including themselves. This kind of tension between current reality and envisaged future indicates the direction for people's thoughts and actions regarding organizational resources in the future. Figure 3.24 illustrates that stakeholders perceive organizational resources through ontologies, and that perception is collected for management purposes for decision making.

Instances capture people's tacit knowledge related to concepts, since the human mind is the best place to understand and interpret the real world and how it changes based on one's personal experience. Catching the current trends and issues from the bottom up gives managers a great opportunity to manage the right concepts and constructs. This way a manager can gain ingenious insights with which he or she can manage the organization. An instance combines several different elements together; cf. the holistic concept of man metaphor by Rauhala (1986). Table 3.12 shows the attributes of an instance.

Involved stakeholders can evaluate the current reality and the future vision of their working surroundings according to the indicators in the Evolute system. The difference between the self-perceived current reality and the vision, objectively speaking, is called creative tension (Senge 1990). In the case of an external object, the difference between the perceived current reality and future vision can also be called proactive vision (Vanharanta et al. 2012). The Evolute system computes the meaning of perceptions based on ontology, Figures 3.25 and 3.26. Figure 3.25 illustrates the Deltoid tool–operator competences (Nurminen 2003).

On the left-hand side in Figure 3.25, we can see the concepts in the ontology. On the right-hand side, we can see the computed meaning of one's perception. The upper darker bars show current reality and the lower lighter bars show envisaged future. The concepts are sorted according to the highest creative tension from the

TABLE 3.12
Class View with the Attributes of an Instance Showing How the Dataset Is Specified

Ontology
- Identifier
- Version
- Concepts, heuristics
- Structure, constructs, grouping
- Indicators

Self-evaluation
- Person ID/anonymous
- Stakeholder group/other grouping
- Demographics
- Timestamp
- Duration
- Reflection of the evaluation

Perception/meaning
- Input values for indicators/raw data
  - Current reality
  - Envisioned future
- Computed meaning for concepts/concept groups
  - Current reality
  - Envisioned future
  - Absolute creative tension (target/current)
  - Relative creative tension (target/current)
- Importance/personal context

top down. Conceptual thinking and objectivity have the highest creative tension in this case. The difference between target and current is the creative tension or proactive vision that is especially interesting for the management since it shows the development needs of the organizational resource as communicated from the bottom up. The absolute creative tension is the target–current, and the relative creative tension is the target/current, that is, the Evolute index. Table 3.13 illustrates their difference. A and B are sample self-evaluations by two different individuals.

In Table 3.13, we can see that A and B have the same creative tension even though A has much higher C and T values than B. However, B has a much higher Evolute index than A. The Evolute index has the benefit of somewhat removing the differences of personal levels in self-evaluations. Some people tend to evaluate things at a high level, while others are more modest in their evaluation. One of the experiences has been that beginners typically tend to evaluate current values higher, and more experienced professionals tend to evaluate current values lower. A typical example has been that some students' current reality is so highly evaluated that there is not much room for improvement. At the same time, an internationally known expert

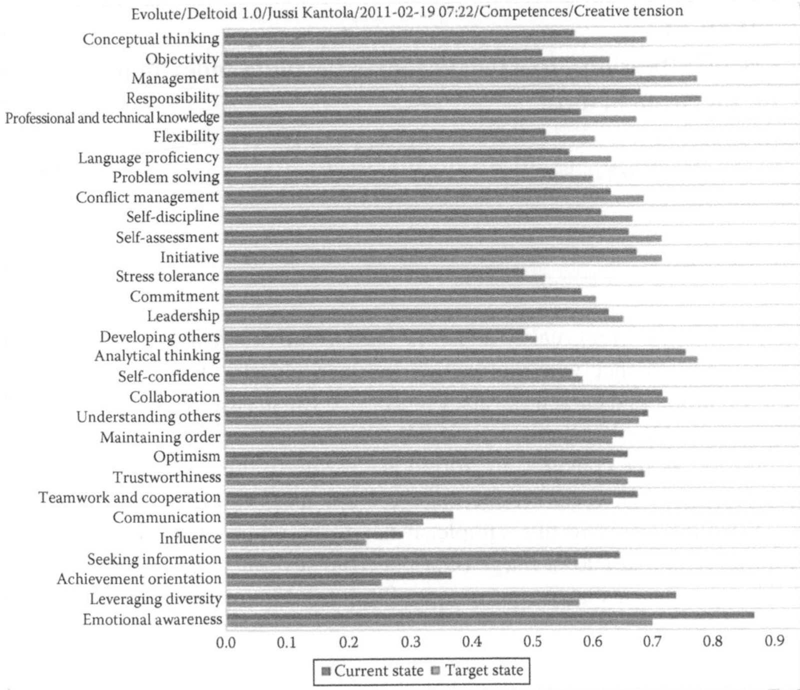

FIGURE 3.25 A sample computed meaning of one's competency based on the Deltoid tool. (From Nurminen, K., Deltoid—The competencies of nuclear power plant operators, MSc thesis, Tampere University of Technology, Pori, Finland, 2003.)

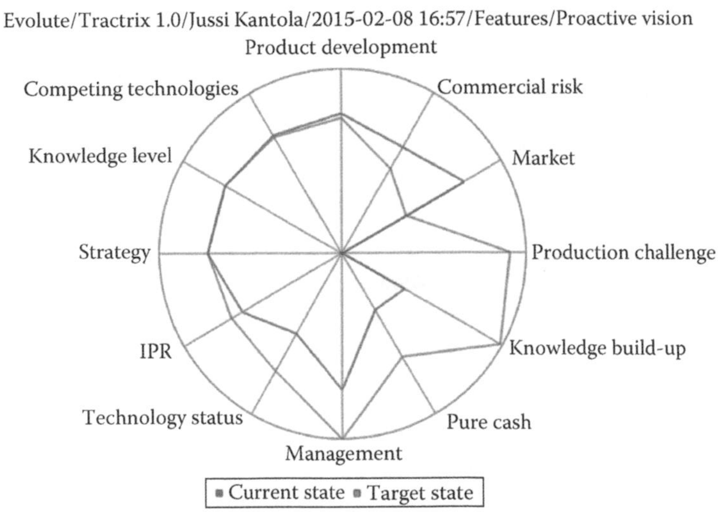

FIGURE 3.26 A sample computed meaning of an R&D project portfolio based on the Tractrix tool (From Naukkarinen, O. et al., A new qualitative decision support tool for evaluating and managing R&D investments, in *Proceedings of EURAM 2014*, St. Andrews, U.K., 2004.)

TABLE 3.13
Creative Tension/Proactive Vision and Evolute Index Illustrated on the Scale of 0–10

	Current (C)	Future (T)	Creative Tension/ Proactive Vision (T–C)	Evolute Index (T/C)
A	8	10	2	1.2
B	2	4	2	2

professor had low current reality and very high creative tension. Figure 3.26 shows another sample instance where proactive vision is clearly visible as the difference between the target and current levels. Figure 3.26 illustrates the Tractrix tool–R&D project portfolio (Naukkarinen et al. 2004).

In Figure 3.26, the concepts are sorted according to the highest proactive vision (starting at 3 pm clockwise). Production challenge and knowledge buildup show the highest proactive vision in this sample. It has to be emphasized that instances are truly unique (Kantola 2005). Large datasets from several years of research support the observation that instances are unique. Figure 3.27 emphasizes this observation

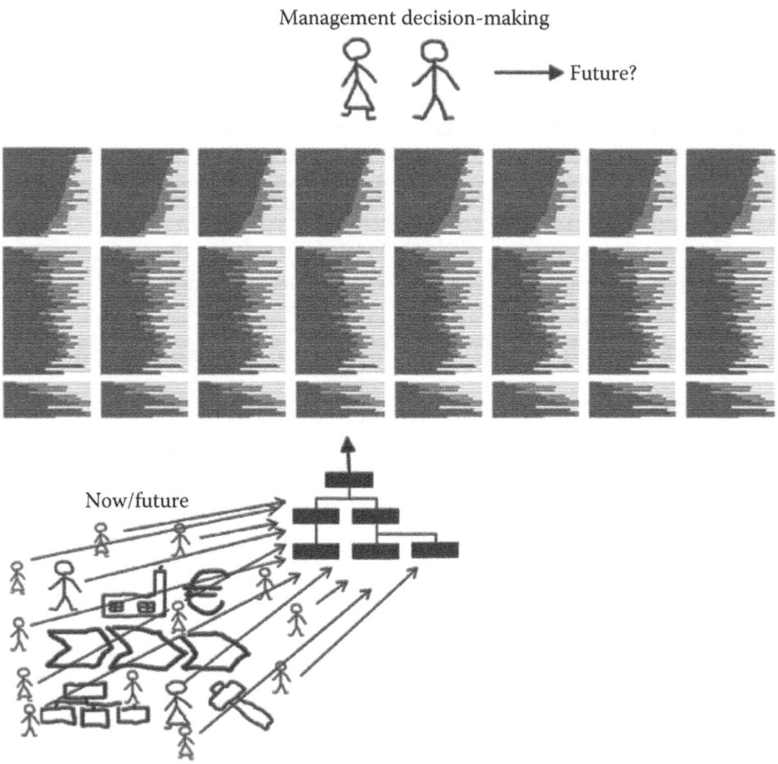

FIGURE 3.27 Does cloning work well for management?

in a case where there are three organizational resources under the focus of management. In recognizing the uniqueness of instances, there is great potential for management to utilize their organization's knowledge assets!

Ontology can be described as $\text{ONTOLOGY}_{\text{Identifier}}$ where ONTOLOGY or O for short stands for the name or code of the ontology model, for example, $O_{\text{Cardioid 1.0}}$ where cardioid is the name of the ontology and 1.0 is the version of the ontology. The specific instance can be described as $O_{\text{Identifier}}$ (Individual, Dataset) where *Individual* is the person identifier and *Dataset* identifies the specific dataset by a sequence number or time stamp. For example, Jack perceives the company's R&D project portfolio using Tractrix tool (Naukkarinen et al. 2004), and the Evolute system computes the meaning of Jack's perception $O_{\text{Tractrix 1.0}}$ (123, 1), where 123 stands for the person ID of Jack and 1 is the sequence number of the first self-evaluation.

The instance is the basic unit of the management of organizational resources according to the Evolute approach. Even though the instance captures several abstract elements, the instance itself is an explicit dataset that is further computable and usable. However, one instance is not enough to manage organizational resources. One instance is like a snapshot. There is more than one organizational resource *at use* in any given moment. There is also more than one person involved at work—typically several stakeholder groups, each having more than one person. Time is one more dimension to consider since longitudinal observation of organizational resources is crucial to understand them. Therefore, for meaningful management purposes, more instances are needed horizontally (many resources), vertically (many people), and longitudinally (dynamics). Figure 3.28 shows a simple dashboard for individuals to create new instances and manage unfinished and existing ones.

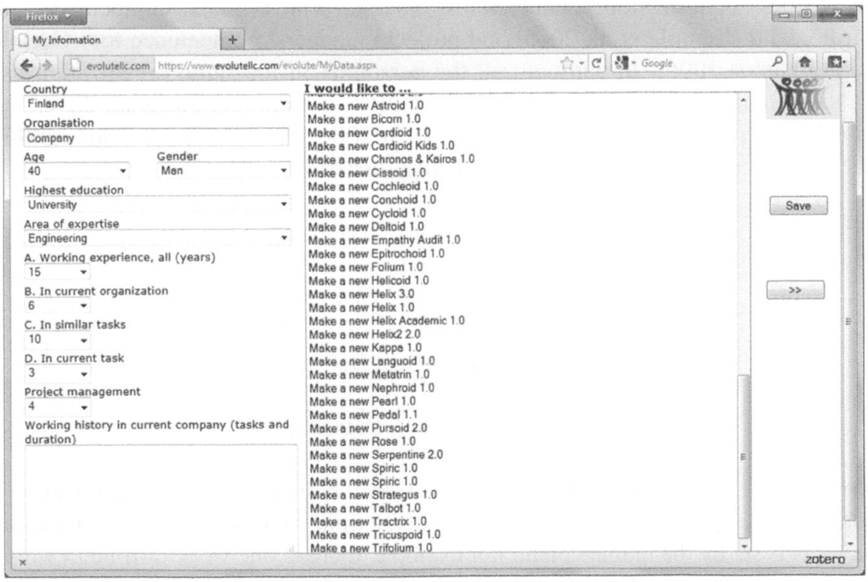

FIGURE 3.28 Individuals can manage their instances in the Evolute system.

The Evolute system supports the following main functions from an individual point of view: (1) signing up and signing in oneself, by invitation, by project code, or by alias; (2) creating instances through self-evaluation, new, continuing, or unfinished; and (3) viewing and using instances. In the next sections we look at the collections of more than one instance.

3.4.2 Instance Vector

The collection of one ontology and more than one individual forms a vertical instance vector and can be described as $O_{\text{Identifier}}$ (Individuals$_{1-n}$, Dataset$_1$). For example, 25 maintenance persons of a company who use the safety culture tool Serpentine 2.0 (Salo 2008, Porkka and Paajanen 2014) once can be described as $O_{\text{Serpentine 2.0}}$ (Person$_{1-25}$, Dataset$_1$). Instance vector is useful for management when there is a special focus on certain organizational resources at a specific moment in time. It gives a snapshot of the current and future views of the organizational resource of many people. It shows the current reality as observed by many and suggests which concepts of the organizational resource are in good shape and which concepts need the most attention. However, typically there are many organizational resources to be managed at once. This means that we need an instance vector of each important organizational resource. This kind of horizontal knowledge expansion can be done with the help of the instance matrix (Kantola 2009) that is described in Section 3.4.3.

3.4.3 Instance Matrix

The collection of many instance vectors forms an instance matrix that can be described as $O_{\text{Identifiers 1-m}}$ (Person$_{1-n}$, Dataset$_1$). For example, all personnel in a project business organization conduct self-evaluation with physical competence tool Cardioid 1.0 (Kantola et al. 2005) and project managers' competence tool Cycloid 1.0 (Liikamaa 2006). In addition, all the personnel in the same organization use the safety culture tool Serpentine 2.0 (Salo 2008; Porkka and Paajanen 2014) and knowledge creation tool Folium 1.0 (Paajanen 2006; Paajanen et al. 2004, 2006). The corresponding instance matrix looks like this:

$$\{O_{\text{Cardioid 1.0}}\,(\text{Person}_{1-n}, \text{Dataset}_1),$$

$$O_{\text{Cycloid 1.0}}\,(\text{Person}_{1-n}, \text{Dataset}_1),$$

$$O_{\text{Serpentine 2.0}}\,(\text{Person}_{1-n}, \text{Dataset}_1),$$

$$O_{\text{Folium 1.0}}\,(\text{Person}_{1-n}, \text{Dataset}_1)\}$$

The instance matrix provides a collective and holistic view to organizations' resources at a specific point in time. Typically, a time window for joining and using the tools is set by the management. For example, there is one week of time in which to participate in the data collection and use the tools during February 1–7. In this case, the Dataset$_1$ would refer to all those instances generated during February 1–7.

The instance matrix is still a snapshot at one point in time. In Section 3.4.4, the longitudinal approach to instance matrices is explained.

3.4.4 Dynamic Instance Matrix

The instance matrix, as a function of time, makes the data collection a longitudinal activity where persons provide more than one instance with the same tool over a longer period of time. Such longitudinal datasets tell the perceived holistic story of the resources of the organization over some time period. In other words, it is a longitudinal story of the organization's assets over time. It also enables the validation of previous decisions and actions, since instances capture current view and future view. Therefore, the future view (envisaged future) of the $\text{Dataset}_1 \approx$ the current view (what came true) of the Dataset_2. It is not of course exactly like this, but that is the rough idea. The dynamic instance matrix can be described as $\text{ONTOLOGY}_{\text{Identifiers 1-m}}$ (Individuals_{1-n}, Dataset_{1-k}). Using the previous example of the instance matrix, the dynamic instance matrix with seven time periods can be expressed as

$$\{O_{\text{Cardioid 1.0}}\ (\text{Person}_{1-n}, \text{Dataset}_{1-7}),$$

$$O_{\text{Cycloid 1.0}}\ (\text{Person}_{1-n}, \text{Dataset}_{1-7}),$$

$$O_{\text{Serpentine 2.0}}\ (\text{Person}_{1-n}, \text{Dataset}_{1-7}),$$

$$O_{\text{Folium 1.0}}\ (\text{Person}_{1-n}, \text{Dataset}_{1-7})\}$$

If the cycle to collect the instance matrix is once every six months, then the aforesaid sample matrix tells the perceived story of the resources of an organization over the time period of three years. Of course, datasets are rarely complete. There can be some variation in those individuals who can participate in the effort. In longitudinal research, a panel study involves a random sample of participants whereas a cohort study aims to keep the participants the same or similar (Menard 1991). Reflective validation of the management decision and action during the next cycle gives a retrospective view of the longitudinal observation of organizational resources.

The content of the instance matrix that is used to manage and develop organizational resources should naturally match those organizational resources that belong to the specific management portfolio in focus. Utilizing systematic ontology-based ways to develop and manage organizational resources enables the managers to have a much larger collection of resources in their management portfolio than is humanly possible. The development and management of organizational resources becomes, step by step, more like a systematic method than an attempt to use experience and tacit knowledge that people have and may share with others. Instance vectors and instance matrices are intermediate steps toward utilizing the dynamic instance matrices over the longer period of time. Then, management can become a continuous process where decisions and actions that have been taken can be validated in the next management cycle from an organizational learning point of view. One *feature* of instances and instance matrices has to be emphasized. Literally, they are instant!

A participant can immediately observe the meaning of his or her perception, and management can immediately see the holistic meaning of collective perceptions. This is a great benefit for speedy decision making and action. There is no need to wait for the results from an analysis by researchers or consultants.

3.4.5 Instance Representation and Visualization

Instances can be looked at on an individual basis and on a group level as instance vectors, instance matrices and dynamic instance matrices. Each individual instance can easily be observed and analyzed in a visual format, where current reality, future vision, absolute creative tension/proactive vision, and relative creative tension/proactive vision are clearly visible without numbers. Each instance or instance group can also be observed and exported in numerical format that allows for further mathematical analyses with external tools.

Grouping instances refers to putting together many instances in the same dataset. The grouping can be based, for example, on demographic data: gender, age, education, tasks at work, etc. Each instance can naturally belong to more than one group simultaneously. Grouping instances from the whole project creates instance matrices for the whole project either in one or more self-evaluation cycles. Instance vectors, instance matrices and dynamic instance matrices are grouped instances. These kinds of dataset can also easily be observed in visual format in many different kinds of graphs, such as bars, spider webs, and lines. One great visual exploratory data analysis method is SOMs by Kohonen (2001) that reveals hidden patterns, clusters and the nature of numerical datasets. Figure 3.29 shows a SOM of 247 Folium instances

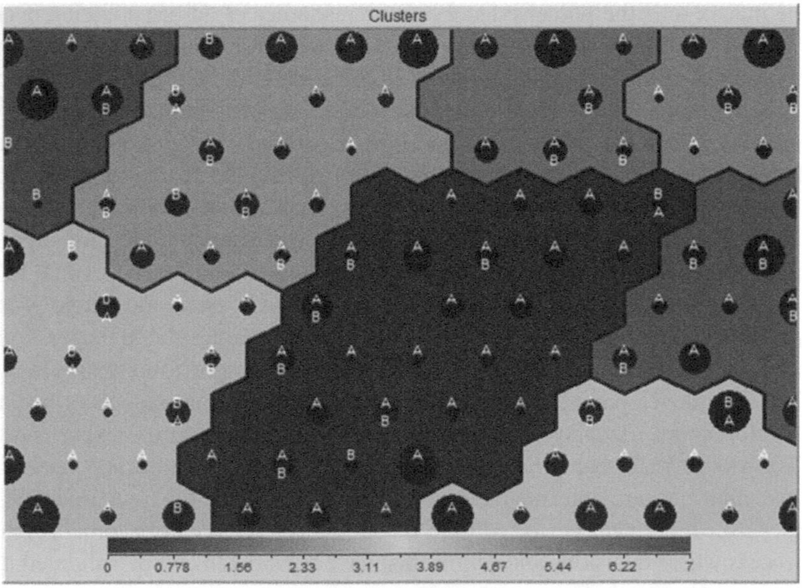

FIGURE 3.29 The SOM of the Folium dataset of 247 instances: A, academic; B, business.

(cf. Kantola et al. 2012). The dataset contains both academic (A) and business (B) instances. The SOM suggests that the dataset has eight instance clusters. The characteristics of each cluster can be quite easily described in normal language by looking at the low values, high values, and proactive vision/creative tension of the concepts in each cluster. These values can be examined on the feature planes of the SOM. In this sample, each cluster has its own characteristics and the decisions can vary by cluster.

Instance matrices can also be observed by the raw data that is the absolute values of inputs (indicators) when participants did the self-evaluation. Raw data can be exported for further analyses with external tools. However, it is important to remember that it is likely to be a waste of time to try to make precise mathematical analyses out of nonprecise datasets, such as instance vectors and matrices. Instance matrices can be observed and analyzed in many nonprecise ways, for example, using ranking and rating methods. In any case, management does not have much use for the precise mathematical results. Section 3.4.6 describes how instance matrices are used for management purposes in real life.

3.4.6 Ingenious Management

Coevolutionary management is still an emerging paradigm (Vanharanta 2005; Vanharanta et al. 2005) that considers coevolution in human and business performance. The goal of the new paradigm is to shift the traditional focus of management activities toward understanding the natural processes of continuously coevolving individuals and the organizations in which these individuals work (Vanharanta 2005; Vanharanta et al. 2005).

Systems coevolve through integration and interaction (Turchin 1999). In reality, organizational resources coexist and therefore need to be counderstood and developed in a coevolutionary manner. By taking an active and promoting role in the coevolution of organizational resources, the Evolute approach aims at a better management of organizational resources than was possible before, and thus the faster learning and development of an organization. Management that utilizes the Evolute approach can make this coevolution occur much faster than it normally would: model organizational resources as ontologies (Proposition 1) > involve the right people (Proposition 2) > compute the meaning (Proposition 3) > manage in the context (Proposition 4). We can see that by making these propositions true in the organization, the management can really make organizational resource systems coevolve faster in real life. That means that individuals and organizations can coevolve faster than otherwise would be possible.

The decision making based on meaning is a step forward, not back. This is illustrated in the following two figures. Figure 3.30 shows how the precisiation of the meaning advances from indicators to instances, and further to concept-based decision making and action (A). This is a step forward in which the increased understanding and meaning is used to proceed in developing organizational resources. The other direction in Figure 3.30 shows an indicator-based decision making and action (B). That is not a way forward since indicators are perceivable evidences of concepts (cf. phenomena by Gillette [2000]). We are interested in the concepts, not their evidences, in order to make decisions that will lead to permanent changes.

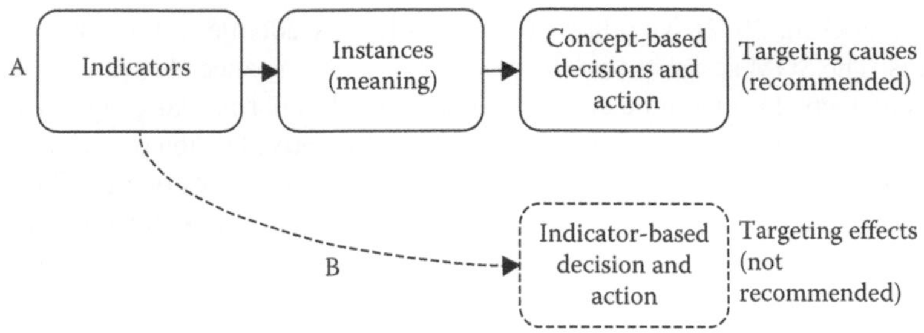

FIGURE 3.30 Targeting causes (A) versus targeting effects (B) in decision-making.

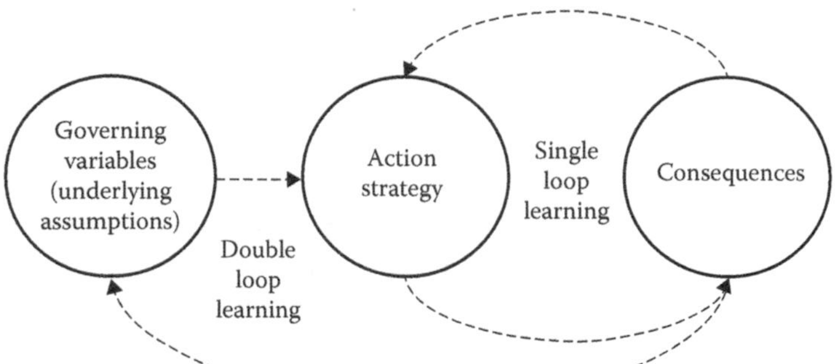

FIGURE 3.31 Single- and double-loop learning in the context of organizational resource management.

Meaning-based decision making is Ingenious Management, and indicator-based decision making is not. There is nothing ingenious in making decisions and actions based on indicators that are the effects of causes. Figure 3.31 illustrates the single and double loop learning by Argyris and Schön (1978). It is clear that indicator-based decision making and action will not provide long-term improvement since it basically aims at fixing effects. But decision making and action that will affect underlying concepts and related assumptions, that is, causes in organizational resource domains, in turn, aim to provide longer-term improvements.

The goal is to make meaning-based decisions to improve the underlying assumptions and ways in which organizational resources function and are set up. The output of effective planning is an ever-changing plan reflecting the continuous learning and adaptation of those who prepare it (Ackoff 1986). According to Argyris and Schön (1996), organizations learn through a continuous cycle: organization's actions → feedback systems → the interpretation of organization's shared knowledge (vision, strategy, and goals) → the development of organization's mental models, actions, and know-how → organization's actions. Therefore, management must be realized in short enough periodic cycles. Figure 3.32 shows a cyclic view of the Evolute approach.

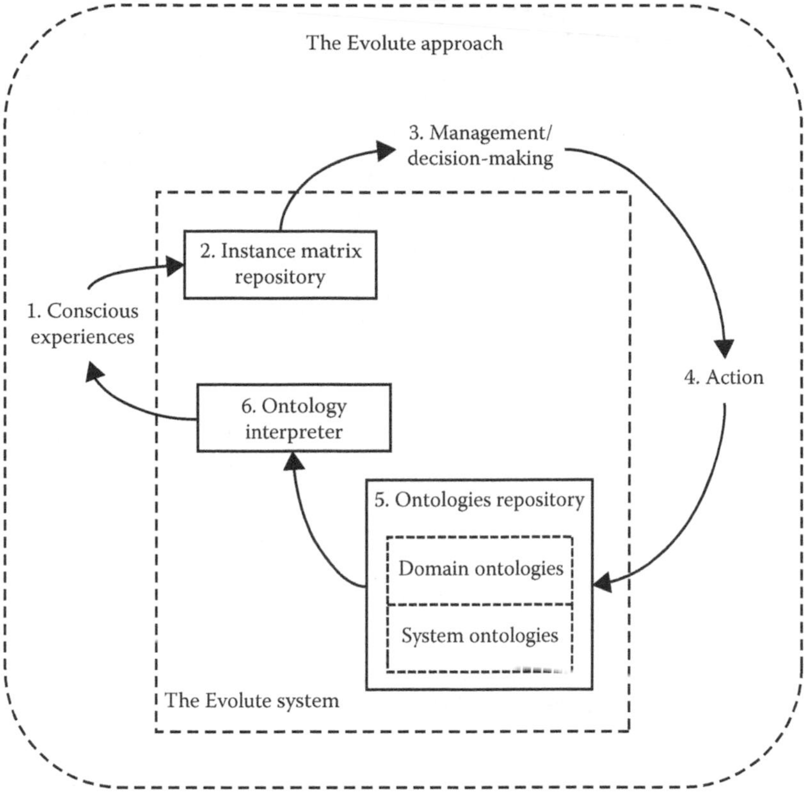

FIGURE 3.32 The Evolute approach combines individuals, ontologies, and the Evolute computing system.

The steps in the cyclic process can be described as follows:

1. Stakeholders perceive organizational resource ontologies through indicators in their conscious experience within a given time window.
2. Instances are stored in an instance repository.
3. Management makes decisions. New action plans are made. The impact of the actions resulting from previous cycles is validated or viewed in a reflective manner. Targeted action plans are made based on credible knowledge input:
   a. Unique instances
   b. Strategy and focus areas of the organization
   c. Relevant explicit knowledge available about the organizational resource, if any
4. Actions are carried out.
5. Accumulated experience enables adjustment of ontology content, presentation, and computation in the ontology repository.
6. Ontology interpreter presents organizational resource ontologies to stakeholders. Go to Step 1.

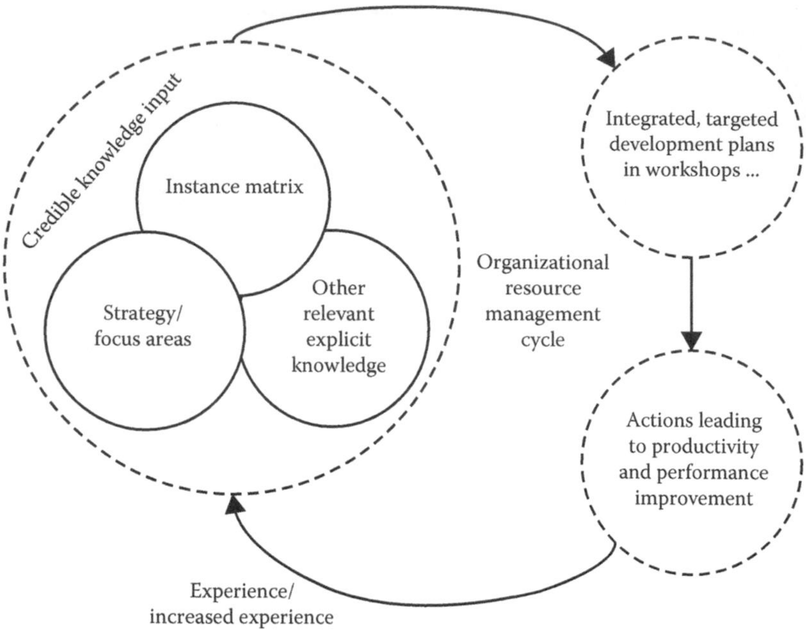

FIGURE 3.33 Organizational resource management cycle or *Ingenious Management* brings the content into the context in periodic cycles.

Figure 3.33 illustrates the same process as in Figure 3.32 but emphasizes the third and fourth steps of the cycle where targeted action plans are made in workshops (these have been working out well). There are three elements that the workshop participants can utilize to come up with targeted action plans. The first element is instance vectors/matrices. The second element is the strategy and focus areas of the organization. This element is typically presented by the management that is participating in the workshop. The third element is any relevant explicit knowledge available about the organizational resource to help with decision making.

Continuous planning requires a continuous process. In the previous academic work, this process was called the Ingenious Management process. It is basically the name of the process aspect of the Evolute approach. According to this process, all organizational resources can be managed through these same steps regardless of their content. The steps are generic and are independent of the conceptual content and structure of organizational resources. Small or large, simple or complex, the same steps still work according to the Evolute approach. Every cycle can be seen as different, and all cycles can, in principle, be handled in a uniform way. It is clear that if some shortcuts are made in the process by skipping step(s) the Ingenious Management process and the Evolute approach will not work.

Additionally, it has been observed that those people who became involved in the approach showed that they were motivated in participating in the development effort. The author would like to emphasize the importance of involving the stakeholders, who have joined the process already in earlier steps, in the workshops. They have been highly motivated in planning for the action.

Now that we recall the whole process we can describe it in another way based on the precisiation term coined by Lotfi Zadeh in connection with the precisiation of meaning (Zadeh 2005). The Evolute approach can also be described as a three-level precisiation process as follows:

1. Perception (with the help of indicators)
2. Meaning (with the help of ontologies and fuzzy logic)
3. Situational management decisions (instance matrices, strategy, and explicit material)

These three levels of precisiation are needed for the current and future positioning of organizational resources, that is, for the management of organizational resources.

The Evolute approach described in this book can be used to develop and manage traditionally difficult concepts in practice and at the same time to involve people in an organization in the effort. The approach has proven its worth in a large number of real-world cases in several counties, in many different kinds of organization, and with many different languages.

From an organization point of view, the Evolute system supports the main functions of (1) starting new instance projects, (2) managing existing projects, and (3) viewing and exporting collective results (not individual results). Figure 3.34 shows the simple manager's dashboard to manage instance projects. Project and person IDs are omitted in the figure.

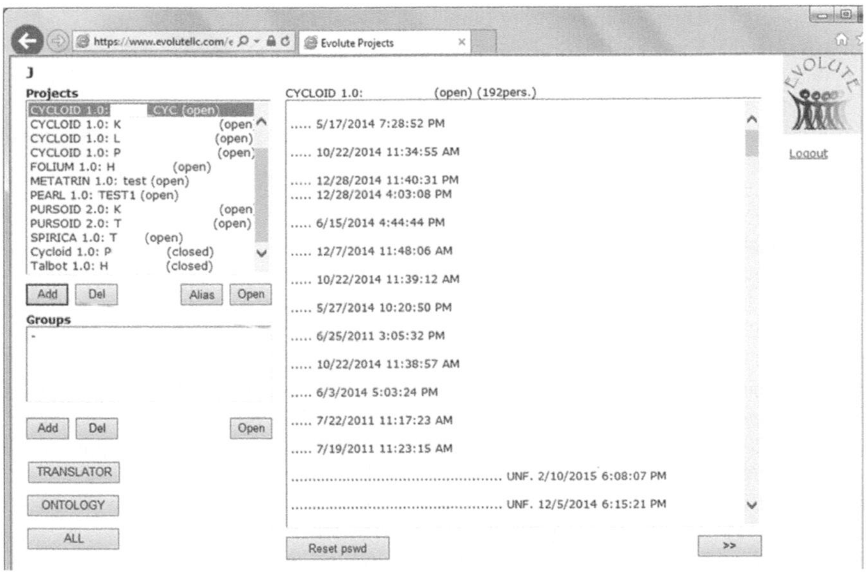

FIGURE 3.34 View for managing instance projects.

TABLE 3.14
Four Cases Describe Different Tools and Their Use according to Evolute Approach

Name of the Tool and Ontology	Current Version	Domain
Pursoid (by Hannu Vanharanta), Chapter 4	2.0	Management of innovation competence in an organization
Helix (by Jarno Einolander), Chapter 5	3.0	Management of commitment in an organization
Kappa (by Kari Ingman), Chapter 6	1.0	Management of sales culture in an organization
Chronos and Kairos (by Tero Reunanen), Chapter 7	1.0	Management of time in an organization

Chapters 4 through 7 that follow describe four cases of the Evolute approach. The organizational resource ontology models have been developed and used in the cases according to the Evolute approach as described in this chapter. The cases are listed in Table 3.14.

REFERENCES

Ackoff, R.L. 1986. *Management in Small Doses*. New York: John Wiley & Sons Inc.

Anttila, R. 2013. Ostotoiminnan semanttinen ontologia (Semantic ontology of buying). MSc thesis, Vaasa, Finland: University of Vaasa, 91pp.

Aramo-Immonen, H. 2009. Project management ontology—The organizational learning perspective. Doctoral thesis, Pori, Finland: Tampere University of Technology.

Argyris, C. and Schön, D.A. 1978. *Organizational Learning*. Reading, MA: Addison-Wesley.

Argyris, C. and Schön, D.A. 1996. *Organizational Learning II: Theory, Method and Practice*, 305pp. Reading, MA: Addison-Wesley.

Beardwell, I. and Holden, L. 1995. *Human Resource Management: A Contemporary Perspective*. London, U.K.: Pitman Publishing.

Berkan, R.C. and Trubatch, S.L. 1997. *Fuzzy Systems Design Principles—Building Fuzzy IF-THEN Rule Bases*. New York: Wiley-IEEE Press.

Berners-Lee, T., Hendler, J., and Lassila, O. 2001. The semantic web. *Scientific American* 284(5): 34–43.

Cronbach, L.J. 1990. *Essentials of Psychological Testing*, 5th edn. New York: Harper/Collins.

Davenport, T.H. and Prusak, L. 2000. *Working Knowledge—How Organizations Manage What They Know*. Boston, MA: Harvard Business School Press.

Davies, J., Fensel, D., and Hamelen, F.V. 2003. *Towards the Semantic Web: Ontology-Driven Knowledge Management*. Chichester, U.K.: John Wiley & Sons Ltd.

Dessler, G. 2001. *A Framework for Human Resource Management*, 2nd edn. Upper Saddle River, NJ: Prentice Hall.

Edgington, T., Choi, B., Henson, B., Raghu, T.S., and Vinze, A. 2004. Adopting ontology to facilitate knowledge sharing. *Communications of the ACM* 47(11): 85–90.

Einolander, J. and Vanharanta, H. Degree of commitment among students at a technological university—testing a new research instrument, In Ahram, T., Karwowski, W., and Marek, T. (eds.), *Proceedings of the 5th International Conference on Applied Human Factors and Ergonomics*, AHFE 2014, July 19–23, 2014, Krakow, Poland, pp. 1767–1778.

Gillette, J.E. 2000. Information is knowledge in motion: A practical framework for understanding knowledge management. Muncie, IN: Ball State University.

Gomez-Perez, A. 2004. Ontology evaluation. In Staab, S. and Studer, R. (eds.), *Handbook on Ontologies*, pp. 251–273. Berlin, Germany: Springer-Verlag.

Gruber, T.R. April 1993. A translation approach to portable ontologies. *Knowledge Acquisition* 5(2): 199–220.

Haaramo, A. 2007. Ostopäällikön kompetenssit (Buyer's competences). MSc thesis, Pori, Finland: Tampere University of Technology.

Halima, T. 2007. Safety culture ontology—From theory to practice. Licentiate thesis, Pori, Finland: Tampere University of Technology.

Harvey-Jones, J. 1986. *Making It Happen: Reflections on Leadership*. London, U.K.: HarperCollins Publishers Ltd.

Heikkilä, A. 2005. Linkage off supply and value chain analysis to the decision support system in a E-business context. MSc thesis, Pori, Finland: Tampere University of Technology.

Hurme-Vuorela, E. 2004. The self evaluation application for the competencies of human work. MSc thesis, Pori, Finland: Tampere University of Technology.

Jusi, J. 2010. Service culture. MSc thesis, Pori, Finland: Tampere University of Technology.

Kantola, J. 1998. *A Fuzzy Logic Based Tool for the Evaluation of Computer Integrated Manufacturing, Organization and People System Design*, 183pp. Louisville, KY: University of Louisville.

Kantola, J. 2005. Ingenious management. Doctoral thesis, Pori, Finland: Tampere University of Technology.

Kantola, J. 2009. Ontology-based resource management. *Human Factors and Ergonomics in Manufacturing & Service Industries* 19(6: Special Issue on Managing Real World Concepts): 515–527.

Kantola, J. and Karwowski, W. (eds.) 2012. *Knowledge Service Engineering Handbook*, 599pp. Boca Raton, FL: CRC Press.

Kantola, J., Karwowski, W., and Vanharanta, H. 2005. Creative tension in occupational work roles: A dualistic view of human competence management methodology based on soft computing. *Ergonomia IJE & HF* 27(4): 273–286.

Kantola, J., Paajanen, P., Piirto, P., and Vanharanta, H. 2012. Showing asymmetries in knowledge creation and learning through proactive vision. *Theoretical Issues in Ergonomics Science* 13(5): 570–585.

Kivelä, J. 2011. Kasvuyrityksen organisaatiokulttuuri (Organizational culture of a growth company). Doctoral thesis, Pori, Finland: Tampere University of Technology.

Klein, M., Ding, Y., Fensel, D., and Omelayenko, B. 2003. Ontology management: Storing, aligning and maintaining ontologies. In Davies, J., Fensel, D., and van Harmelen, F. (eds.), *Towards the Semantic Web: Ontology-Driven Knowledge Management*, pp. 47–69. West Sussex, U.K.: John Wiley & Sons.

Klir, J.G. and Yuan, B. 1995. *Fuzzy Sets and Fuzzy Logic: Theory and Applications*. Upper Saddle River, NJ: Prentice Hall, Inc.

Kohonen, T. 2001. *Self-Organizing Maps*. HUT, Finland: Springer Verlag.

Kosko, B. 1994. *Fuzzy Thinking*. London, U.K.: Flamingo.

Leonard, D. and Swap, W. September 2004. *Deep Smarts*, pp. 88–97. Boston, MA: Harvard Business Press.

Liikamaa, K. 2006. Tacit knowledge and project managers' competencies. Doctoral thesis, Pori, Finland: Tampere University of Technology.

Lin, C.T. and Lee, C.S.G. 1996. *Neural Fuzzy Systems—A Neuro-Fuzzy Synergism*. Upper Saddle River, NJ: Prentice Hall, Inc.

Lintula, E. 2004. Epitrochoid—The competencies evaluation method for human resource managers. MSc thesis, Pori, Finland: Tampere University of Technology.

Mamdani, E.H. and Assilian, S. January 1975. An experiment in linguistic synthesis with a fuzzy logic controller. *International Journal of Man-Machine Studies* 7(1): 1–13.

Mäkinen, S. 2004. The strategic management competencies. MSc thesis, Pori, Finland: Tampere University of Technology.

Mäkiniemi, P. 2004. Conchoid—Kunnossapidon Henkilöstön Kompetenssien Itsearviointijärjestelmä (Maintenance personnel's competence self-evaluation system). MSc thesis, Pori, Finland: Tampere University of Technology.

Menard, S. 1991. *Longitudinal Research: Sage University Paper Series: Quantitative Applications in the Social Sciences.* Newbury Park, CA: SAGE Publications.

Mitra, P. and Wiederhold, G. 2004. An ontology-composition algebra. In Staab, S. and Studer, R. (eds.), *Handbook on Ontologies*, pp. 93–113. Berlin, Germany: Springer-Verlag.

Naukkarinen, O., Kantola, J., and Vanharanta, H. 2004. A new qualitative decision support tool for evaluating and managing R&D investments. In *Proceedings of EURAM 2014*, St. Andrews, U.K.

Nonaka, I., Toyama, R., and Konno, N. February 2000. SECI, Ba and leadership: A unified model of dynamic knowledge creation. *Long Range Planning* 33(1): 5–34.

Nurminen, K. 2003. Deltoid—The competencies of nuclear power plant operators. MSc thesis, Pori, Finland: Tampere University of Technology.

O'Connor, J. and McDermott, I. 1997. *The Art of Systems Thinking: Essential Skills for Creativity and Problem Solving.* London, U.K.: Thorsons.

Odrakiewicz, P., Kantola, J., and Vanharanta, H. 2009. Competences and competence models. Languoid: Communication competence management and testing in organizations. In Wankel, C. et al. (eds.), *Proceedings of the Fourth Innovation in Management, Cooperating Globally*, Poznan, Poland, May 21–22, 2009, pp. 361–365.

Orbst, L. 2003. *Ontologies for Semantically Interoperable Systems*, pp. 366–369. New Orleans, LA: ACM.

Paajanen, P. 2006. Dynamic ontologies of knowledge creation and learning. Licentiate thesis, Pori, Finland: Tampere University of Technology.

Paajanen, P. 2012. Managing and leading organizational learning and knowledge creation. Doctoral thesis, Pori, Finland: Tampere University of Technology. Publication 1062.

Paajanen, P., Kantola, J., Karwowski, W., and Vanharanta, H. 2004. Applying systems thinking in the evaluation of organizational learning and knowledge creation. In *Proceedings of CITSA 2004*, Orlando, FL.

Paajanen, P., Piirto, A., Kantola, J., and Vanharanta, H. 2006. FOLIUM—An ontology for organizational knowledge creation. Orlando, FL.

Palonen, E. 2005. The competencies evaluation application for entrepreneurs. MSc thesis, Pori, Finland: Tampere University of Technology.

Parry, D. 2004. *A Fuzzy Ontology for Medical Document Retrieval*, pp. 121–126. Dunedin, New Zealand: Australian Computer Society, Inc.

Pedrycz, W. 1989. *Fuzzy Control and Fuzzy Systems.* Somerset, U.K.: Research Studies Press Ltd.

Piirto, A. 2012. Safe operation of nuclear power plants—Is safety culture an adequate management method? Doctoral thesis, Pori, Finland: Tampere University of Technology.

Porkka, P.L. and Paajanen, P., *Comparison of Safety Cultures between the Chemical and Power Industries*, Vol. 36, 2014, pp. 403–408.

Rastogi, P.N. 1995. *Management of Technology and Innovation—Competing Trough Technological Excellence.* New Delhi, India: Sage Publications.

Rauhala, L. 1986. Ihmiskäsitys Ihmistyössä (The concept of human being in helping people). Helsinki, Finland: Gaudeamua.

Reunanen, T. 2013. Leaders' conscious experience towards time. MSc thesis, Pori, Finland: Tampere University of Technology.

Salo, M. 2008. Serpentine–safety culture. MSc thesis, Pori, Finland: Tampere University of Technology.

Saunders, M., Lewis, P., and Thornhill, A. 2007. *Research Methods for Business Students*. Harlow, U.K.: Pearson Education.

Sayyarrodsari, B. and Homaitar, A. 1997. The role of "Hierarchy" in the design of fuzzy logic controllers. *IEEE Transactions on Systems, Man, and Cybernetics—Part B: Cybernetics* 27(1): 108–118.

Segev, A. and Gal, A. June 2008. Enhancing portability with multilingual ontology-based knowledge management. *Decision Support Systems* 45(3): 567–584.

Segev, A. and Quan, Z.S. 2012. Bootstrapping ontologies for web services. *IEEE Transactions on Services Computing* 5(1): 33–44.

Seikola, S. 2013. Pearl—Application for measuring customer's conscious experience. MSc thesis, Pori, Finland: Tampere University of Technology.

Senge, P. 1990. *The Fifth Discipline: The Art & Practice of Learning Organization*. New York: Currency Doubleday.

Silva, C.W. 1995. *Intelligent Control: Fuzzy Logic Applications*. Boca Raton, FL: CRC Press.

Sure, Y., Staab, S., and Studer, R. 2004. On-to-knowledge methodology (OTKM). In Staab, S. and Studer, R. (eds.), *Handbook on Ontologies*, pp. 117–132. Berlin, Germany: Springer-Verlag.

Sure, Y. and Studer, R. 2003. A methodology for ontology-based knowledge management. In Davies, J., Fensel, D., and van Harmelen, F. (eds.), *Towards the Semantic Web: Ontology-Driven Knowledge Management*, pp. 33–45. West Sussex, U.K.: John Wiley & Sons Ltd.

Taipale, M. 2006. Modeling of a self-assessment system utilizing fuzzy logic. MSc thesis, Pori, Finland: Tampere University of Technology.

Torrington, D. and Hall, L. 1991. *Personnel Management—A New Personnel Approach*, 2nd edn. London, U.K.: Prentice Hall.

Tuomainen, M. 2014. Measuring the degree of company democracy culture. MSc thesis, Pori, Finland: Tampere University of Technology.

Turchin, V.F. July 1999. A dialogue on metasystem transition. *World Futures: The Journal of General Evolution* 45: 213–243.

Uschold, M. and Gruninger, M. 1996. Ontologies: Principles, methods and applications. *Knowledge Engineering Review* 11(2): 93–155.

Vanharanta, H. 2005. Co-evolutionary design for human-compatible systems. In Sinay, J. et al. (eds.), *CAES'2005, International Conference on Computer-Aided Ergonomics, Human Factors and Safety: Information Technology, Knowledge Management and Engineering for Enterprise Productivity and Quality of Working Life*, May 25–28, 2005.

Vanharanta, H. and Kantola, J. 2015. Proactive vision for strategy making. In *Sixth International Conference on Applied Human Factors and Ergonomics (AHFE 2015) and the Affiliated Conferences, AHFE 2015*, Las Vegas, NV, July 26–30, 2015.

Vanharanta, H., Kantola, J., and Karwowski, W. 2005. A paradigm of co-evolutionary management: Creative tension and brain-based company development systems. *CD-ROM Proceedings of the 11th International Conference on Human-Computer Interaction—HCI2005*, July 22–27, 2005, Las Vegas, NV, 10 p.

Vanharanta, H., Kantola, J., and Karwowski, W. 2007. Love dimensions on the web. In *Human-Computer Interaction, HCI Applications and Services*, vol. 4553, pp. 1057–1062. *Lecture Notes in Computer Science*. Berlin, Germany: Springer.

Vanharanta, H., Magnusson, C., Ingman, K., Holmbom, A., and Kantola, J. 2012. Strategic knowledge services. In Kantola, J. and Karwowski, W. (eds.), *Knowledge Service Engineering Handbook*, vol. 4057, pp. 527–556. Boca Raton: CRC Press.

Wilhelm, M.R. and Parsaei, H.R. 1991. A fuzzy linguistic approach to implementing a strategy for computer integrated manufacturing. *Fuzzy Sets and Systems* 42(2): 191–204.

Zadeh, L. 1965. Fuzzy sets. *Information and Control* 8(3): 338–353.

Zadeh, L. 1973. Outline of a new approach to the analysis of complex systems and decision processes. *IEEE Transactions on Systems, Man, and Cybernetics* 1(1): 28–44.
Zadeh, L. 1975. The concept of a linguistic variable and its application to approximate reasoning. *Information Sciences* 8(3): 199–249.
Zadeh, L. October 1983. Commonsense knowledge representation based on fuzzy logic. *Computer* 16: 61–65.
Zadeh, L. June 9, 2005. Toward a generalized theory of uncertainty (GTU)—An outline. *Information Sciences* 172(1–2): 1–40. doi:10.1016/j.ins.2005.01.017.
Zadeh, L. 2014. Impact report of fuzzy logic. Berkeley Initiative of Soft-Computing (BISC) Group, University of California, Berkeley, CA.
Zadeh, L.A. 1971. Quantitative fuzzy semantics. *Information Sciences* 3(2): 159–176.

4 Pursoid
Innovation Competence of Human Resources

*Hannu Vanharanta**

ABSTRACT

Many people would like to be innovative, but only some can create something really new. However, innovation characteristics are something that can be improved, and therefore measuring the degree of innovation competences at the individual, group, team, and the whole company level in any organization is crucial and almost mandatory. In this chapter, we present the current status of our research tool, Pursoid, with which we have had the opportunity to measure the innovation competence of individuals. In addition, we show how we have obtained a realistic view of the way the collective degree of innovation competence can be revealed. All the results can be used to improve and develop the overall innovation competence level inside organizations. It is also possible to educate and teach individuals and the entire workforce to see how they can discover new ways and ideas of how to prioritize, rank, analyze, and improve their innovation competence characteristics and the overall collective innovation competence. Companies can survive better when they have active people who can really see the external and internal changes in their company's business world. Inside companies, innovation competence has a special dimension of added value. Therefore, the degree of innovation competence is one of the most important measurements to be considered and analyzed inside organizations. In this research, we show how this kind of competence can be measured indirectly.

4.1 INTRODUCTION

The need and demand for creativity and innovation in organizations is driven by a few basic factors. First, we have solutions that are too old for our current problems, new technologies that enable many new innovations, competitors quickly learning how to imitate past innovations, customers' increasing demand for new innovations, and the fact that innovations can bring superior long-term financial performance as well as many external and internal changes in the business world (cf. Dhillon 2006, pp. 42–43). Likewise, there are some basic characteristics that boost creativity and innovation (cf. Dhillon 2006, pp. 42–43): management's eagerness to take risks, incentives for innovators, the free flow of information, and employees' free

* Poznan University of Technology, Poland.

access to sources of knowledge. In addition, good new ideas and methods are supported, encouraged, and welcomed. However, motivating people to become more innovative tomorrow than they are today is a challenging task for management and leadership. First, we have to understand from the managerial point of view what lies behind innovation competence and what the main concepts within innovation competence are. Second, from the leadership point of view, we have to understand how to perceive the degree of innovation competence, what the company requirements are, and how to motivate the whole company organization to improve and develop their innovation capacity, skills, and capabilities for active usage, that is, to use their innovation competence. Companies need people who can innovate. They also need people who, in times of tough competition, are ready to change and improve their own innovation competence. This is why managers and leaders need tools, that is, management and leadership applications, with which they can obtain information and knowledge quickly about what needs to be done, changed, and improved so that the overall innovation competence with its characteristics can be perceived, interpreted, understood, nurtured, and grown rapidly within the company. A highly responsive environment is also needed for developing and raising these innovation competence characteristics to achieve results (cf. Ford 1992, p. 248). If this does not happen, the company will encounter stagnation and low growth rates, profitability, productivity, and overall performance. On the individual level, the focus is in personal as well as in social innovation competences. When this individual innovation competence meets the organization's responsive work environment, it opens the way for new innovations to emerge. Therefore, this kind of holistic concept might be the best way to give birth to innovations. We can then say that innovation demand meets innovation supply.

4.2 METAPHORS, CONSTRUCTS, CONCEPTS, AND ONTOLOGY CREATION

Our research started from the Holistic Concept of Man metaphor by Rauhala (c.f. Rauhala 1995) as well as the idea that managers and leaders should be supported by means of different executive support systems (cf. Vanharanta 1995). The Holistic Concept of Man metaphor was the basis for another metaphor: the Circles of Mind by Vanharanta (cf. Vanharanta 2003) and the required technology, that is, fuzzy logic, to support the systems engineering view of the targeted competence research (c.f. Kantola 2005). Combining a situation-based view with the inner brain processes gave rise to a comprehensive application development platform. The main construct of the Circles of Mind was opened up and launched for occupational role research and termed brain-based application research. Supporting sciences, a theoretical framework, methodology, and technology were woven together with the Co-Evolute theory and methodology architecture (cf. Kantola et al. 2010). The research was carried out first with human object ontologies and later with business object ontologies, as shown in Figure 4.1. Several applications are mentioned in the figure with names like Cycloid, Conchoid, Helix, Folium, Liitus, etc. One of them is Pursoid, that is, the innovation competence application. Pursoid has its origin as an important

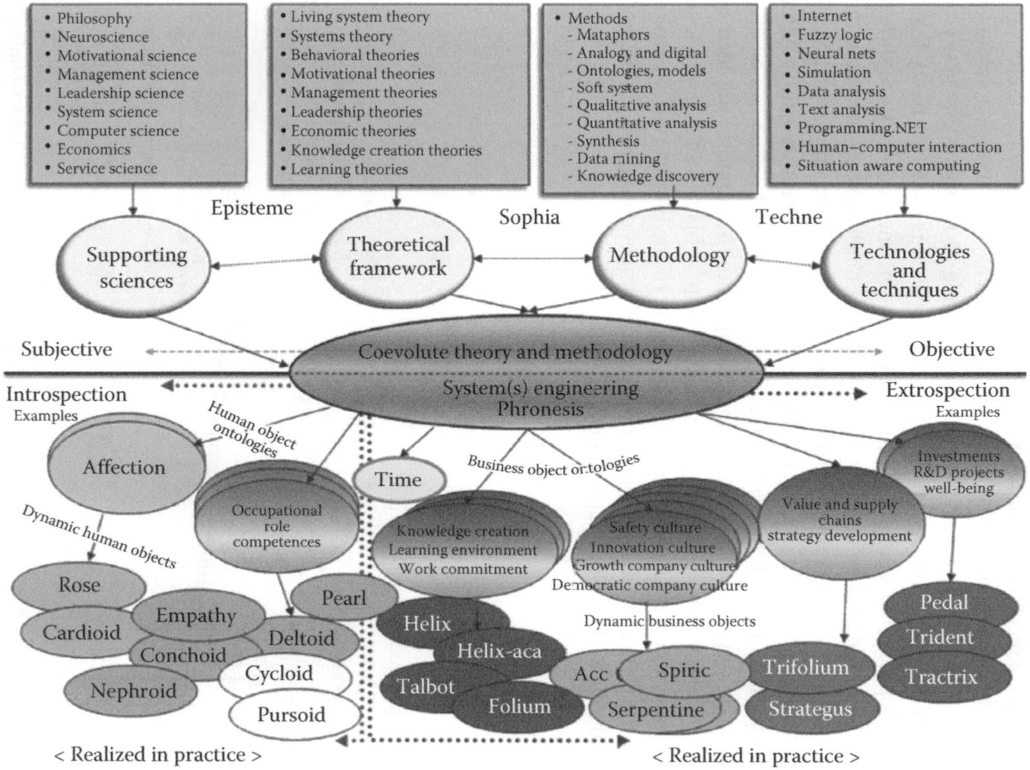

FIGURE 4.1 Coevolute theory and methodology. (Revised from Kantola, J. et al., Managing managerial mosaic: The evolute methodology, in: Ordonez de Pablos, P., Lytras, M., Karwowski, W., and Lee, R. (eds.), *Electronic Globalized Business and Sustainable Development Through IT Management: Strategies and Perspectives*, Business Science Reference, Hershey, PA, 2010, pp. 77–89.)

application for business research and development (R&D), and new product and service creation and development. The innovation competence model is one of the human object ontologies, presented as one of the main occupational competences developed so far as a brain-based research and development application.

The first applications and background were presented back in 2005 in a conference paper by Vanharanta et al. (2005). Situation-aware applications for occupational role competences have been tested in several different occupations: project manager, salesman, executive, social worker, operators, maintenance man, etc., and each of them has been tested in real world circumstances with real test subjects. Many of the ontologies are based on master's theses made at Tampere University of Technology, Pori Campus, Finland, and the technology used is based on fuzzy logic application development (c.f. Kantola 2005). The theory and methodology are shown in Figure 4.2. All of the applications developed with this methodology belong to the area of "situation-aware computing." In the current context, there are several management and leadership ontologies, which are revealed by a literature study. After that, the ontology is examined in detail so that a fuzzy logic application can be created. The methodology, as well as the computer applications, supports the idea of evaluating the ontology in both its current and future state. This way, it is possible to capture the creative tension described by the test subject of the human object ontologies as well as the future proactive vision from business object ontologies (cf. Eklund et al. 2012; Kantola et al. 2006; Paajanen et al. 2006).

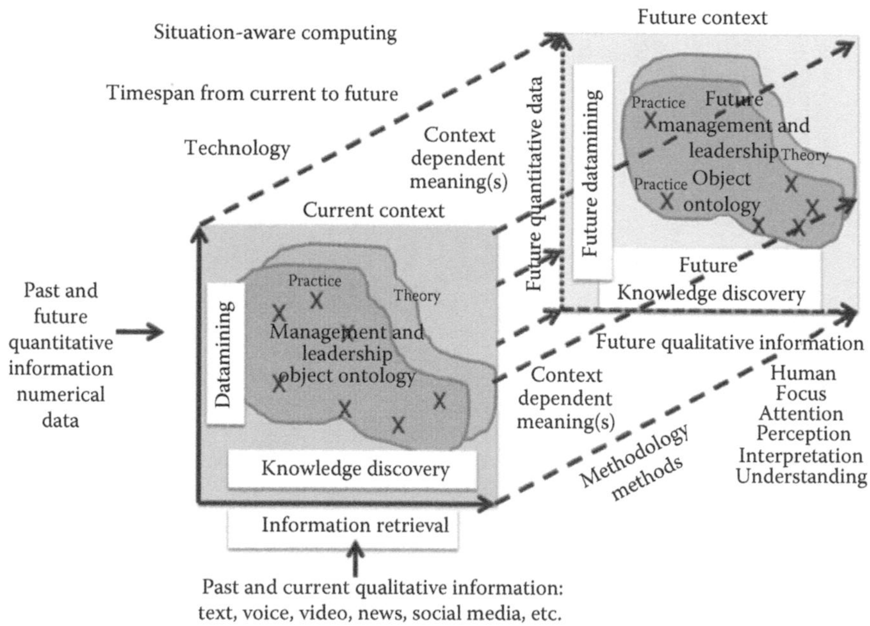

FIGURE 4.2 The construct of situation-aware computing.

4.3 INNOVATION EVALUATION PROCESS

We have created an overall process to understand work competence improvement. In this context, when we look at innovation competence improvement we have to understand that people have different levels of innovation capacity. Some people are not interested in innovating at all; some people feel that everything has already been innovated in their job. However, often people's inner voice (cf. Circles of Mind) tells them that specific aspects need improvement. Our way of thinking starts from the stage where people do their work. They are really the masters and professionals in their work and know a lot about how to improve internal work processes, and how to develop company products and services. They often also get to know what the customers' demands are, and they know how they can add value to the customers' own processes. Figure 4.3 describes the current process we use.

The starting point is the employees who really have their own innovation capacity. This is a hidden capacity to be used in a certain situation. It provides a work-specific innovation capability and, furthermore, when in active use, it provides innovation competence. This competence is also hidden along with its constructs, concepts, and characteristics. By understanding the construct we can emulate the innovation activity in real terms. The construct can have characteristics as well as labels, which can lead us to personal innovation competence. The aim of our research and applications is that the applications are the test and research tools with which we can capture the hidden, that is, tacit, knowledge of innovation competence (cf. Koskinen and

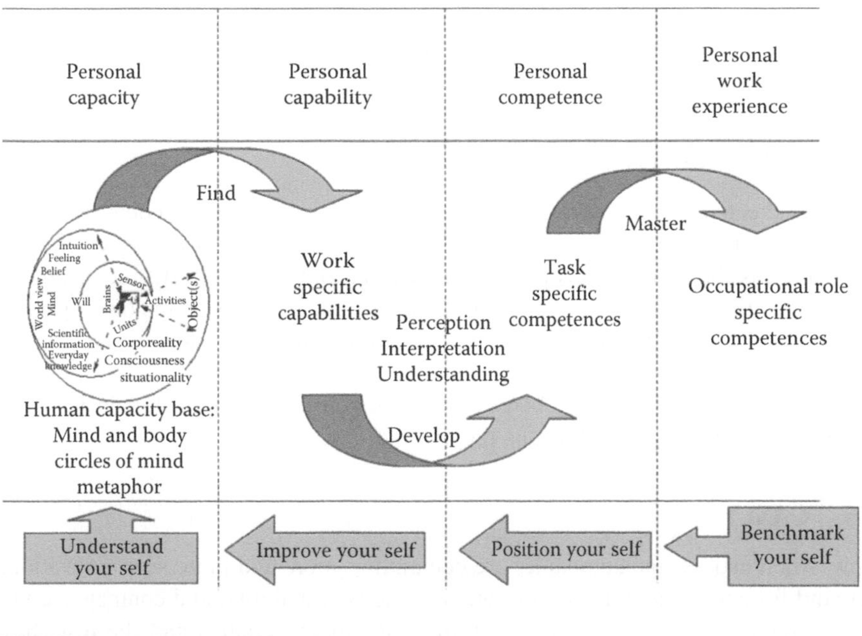

FIGURE 4.3 Work competence improvement process.

TABLE 4.1
Pursoid 2.0 Ontology

Personal competences	Thinking habits	Analytical thinking, divergent thinking, creative thinking, lateral thinking, convergent thinking, intuitive thinking, critical thinking, and strategic thinking
	Attention, experiences	Curiosity, observation, and imagination
	Working habits	Focus and interests, formulating problems, and attitude to work
	Self-awareness	Self-confidence and self-esteem
	Self-regulation	Self-control, trustworthiness, flexibility, responsibility, and stress tolerance
	Motivation	Achievement orientation, autonomy, initiative, change management, and risk taking
	Expertise and development	Occupational and technical expertise, seeking information, absorptive capacity, and self-development
Social competences	Empathy	Understanding others, conflict management, and leveraging diversity
	Relationship management	Communication, networking, teamwork, and cooperation

Vanharanta 2000, 2002). Through personal innovation experience, we can see how our role-specific innovation competence works in practice. The real problem is that people seldom understand the content of innovation competences, and therefore it is difficult for them to improve and develop their innovation capacity, skills, capability, and competences.

4.4 PURSOID TOOL

We have created an innovation competence ontology and have also determined the variables, that is, the concepts as well as the attributes, in this construct. Thus, it has been possible to make a computer application to measure the degree of innovation competence as well as the degrees of various concepts inside the ontology. The ontology created for Pursoid, that is, the innovation competence application, consists of the following main areas and conceptual contents (Table 4.1).

Our thinking behind how to boost innovation culture and innovators is described more in the article by Bikfalvi et al. (2010).

4.5 CASE

Now, when nations and companies are demanding more and more new innovations, demand for expanding human innovation capacity, capability, and competence will be crucial. So far, we have been working only in universities and the tests have been conducted in four different universities. Our test subjects are students who are studying for their master's degrees in industrial economics and engineering. The number of students was a total of 30.

4.6 RESULTS

In the following, we present the collective test results of our test subjects. The results are based on the Pursoid application. The ontology follows the previously mentioned content. The results are presented with three different figures directly from the screen of the application. The collective test results, according the current state, are as shown in Figure 4.4.

The short analysis of the above test results shows that all of the current stage innovation competences seem to be on a very high level. Absorptive capacity is the only characteristic that, in the future state, seems to develop negatively. The results may indicate that absorptive capacity might be difficult to understand in practice. Absorptive capacity has been defined as the ability to exploit external data,

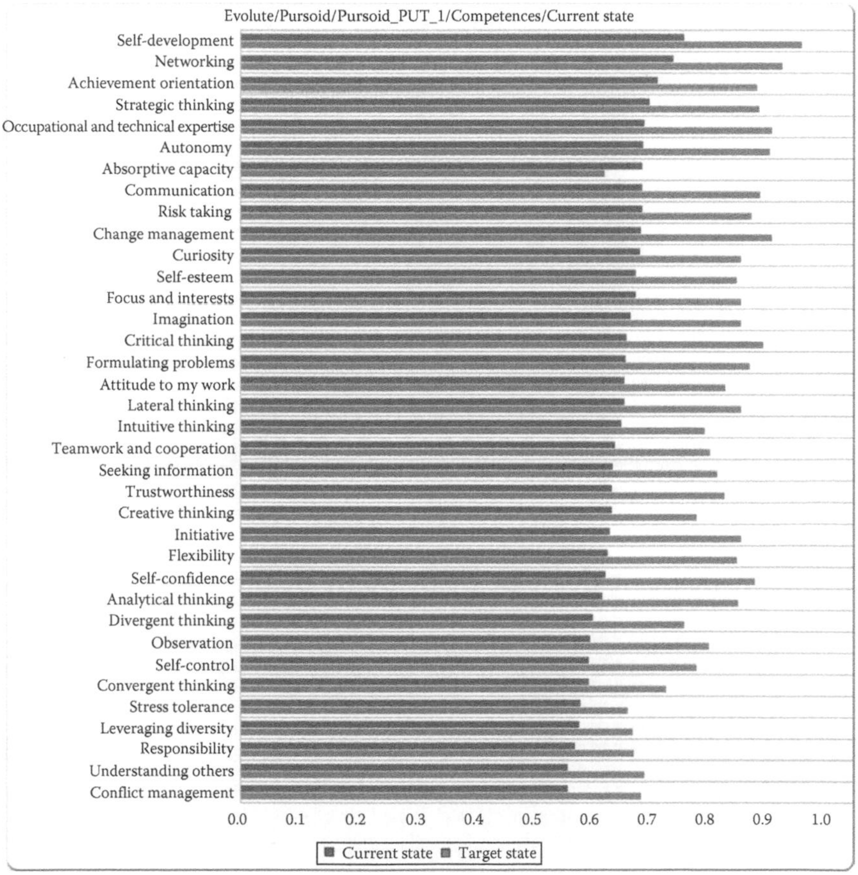

FIGURE 4.4 Pursoid 2.0 Instance sorted by the current state.

information, and knowledge during the innovation process. From the results, leaders and managers can rapidly gain an overall picture of the innovation competences inside their own company. Which characteristics may need improvement? How high do workers rank certain characteristics? What is their creative tension to develop themselves? A company has always its own priorities. What a company prioritizes, and what in the workforce is seen to be important, belongs to the top priorities of managers and executives. The collective test results, according to the future state, are as shown in Figure 4.5.

The short analysis of the test results shown in Figure 4.5 reveals that all the future stage innovation competences seem to be on a very high level. In a numeric form, the results are between 0.62 and 0.97. Absorptive capacity has the lowest figure and the negative development is shown clearly in the bottom of the figure. Self-confidence gets the highest values, which means that people really want to develop their most important own characteristics in their work context. This is really something that leadership must concentrate on and support in their coming

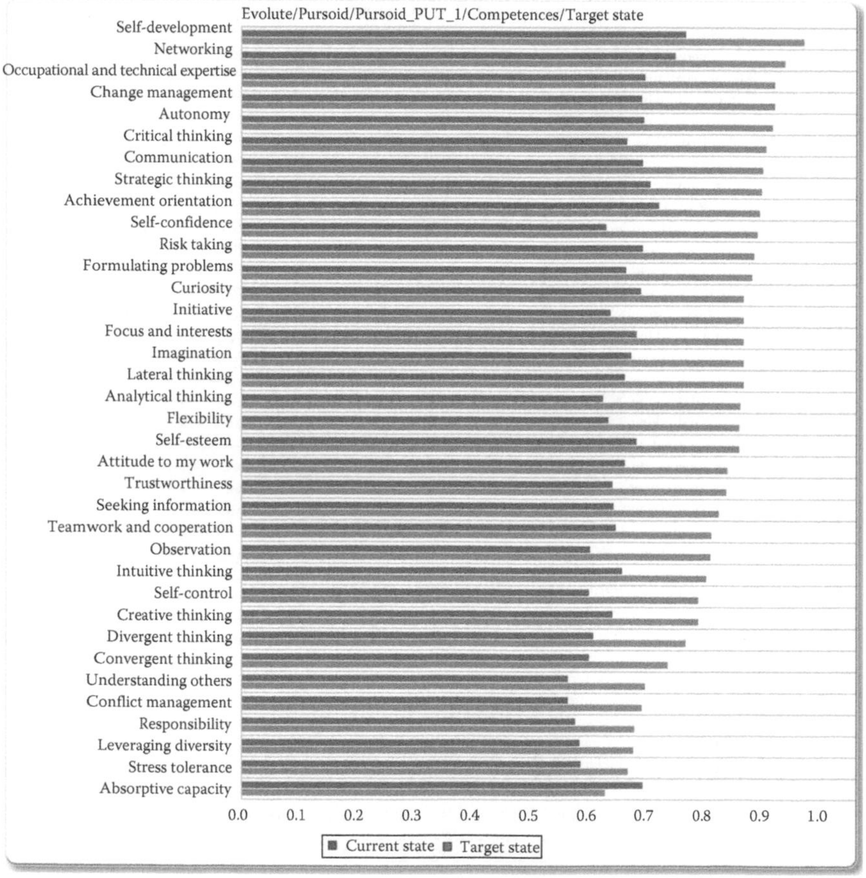

FIGURE 4.5 Pursoid 2.0 Instance sorted by the target state.

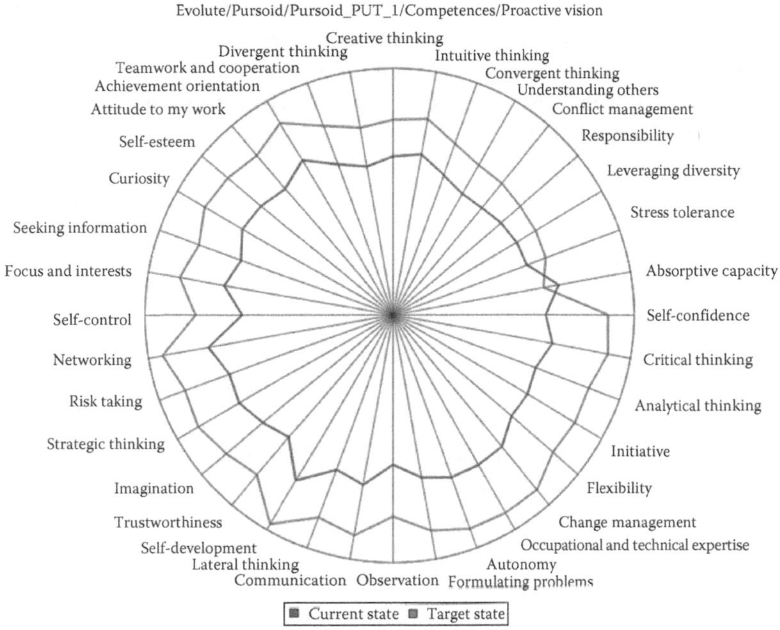

FIGURE 4.6 Pursoid 2.0 Instance sorted by the evolute index (target/current).

decisions. The results also show that people understand very well the importance of networking, occupational, and technical knowledge, change management, and critical thinking. Stress tolerance is low as well as levering diversity. All the differences show clearly that people like to learn more and that management and leadership must do something to help people to better understand the innovation process as well as the characteristics of the innovation competences. The collective test results, according to the evolute index (target/current), are as shown in Figure 4.6.

A short analysis of the results in Figure 4.6 can be concluded as follows: evolute index shows directly the ratio between future state and current stage. The ratio is largest in self-confidence. The result shows also that people are ready to strengthen their self-confidence to a very high level. This in turn means that they are not so satisfied with the current situation. Analytical thinking, initiative, flexibility, and critical thinking get high evolute-index values. From the leadership and management point of view, we can perceive and understand that these collective results clearly support that people are ready to improve their innovation competence, and that many of the competences already have high values. It is easy to see the overall innovation competence with a spiral figure. The reading of the figure starts from the right side of the figure (3 o'clock) (Figure 4.6).

All the innovation characteristics, except absorptive capacity, show a need for development. The creative tension is highest in self-confidence, which is many times the start of personal growth. Critical thinking and analytical thinking, as well as initiative, get very high figures. It is very interesting to see that both the personal as well social competences are scattered throughout the figure. This is why we cannot say directly which of these categories are more important to be developed. From the figure the thinking habits are also scattered, and those that are more linked to creativity are far away from critical thinking and analytical thinking. Thinking habits show, however, that it would be an area to be developed.

4.7 DISCUSSION

Pursoid application is technically operating well. It has been verified with many test runs in three different universities: Tampere University of Technology (TUT), Pori, Finland; University of Vaasa (UVA), Vaasa, Finland; and Poznan University of Technology (PUT), Poznan, Poland. The total number of test subjects is already over 100. The aforementioned results present results with 30 students.

Many positive things exist with this kind of innovation competence application: users of the application can participate in a continuous manner through the internet. Users of the application can use their own language. It is possible to get "collective knowledge" from the current innovation competence situation. Comparisons of different SBUs, SBAs, departments, and their specific situations are easy to create. Results can be ranked by index and by company requirements. The focus can be on target training, target management, and target leadership. Competence applications increase the internal information, knowledge, and understanding to tackle the external complex business environment. These positive things are still under research, and therefore the real validation work has not yet been finalized.

The target is now to widen the innovation competence research to companies. The whole learning loop is also another area to be developed with the test subjects. The content descriptions and the overall ontology are important for understanding the application, as well as for teaching purposes. It is important to note that, before we should develop our skills and talents, it is useful to understand the content of these skills and talents. After this, there is a lot of practical work to enable the innovation competence to grow.

REFERENCES

Bikfalvi, A., Jussila, J., Suominen, A., Vanharanta, H., and Kantola, J. 2010. How to boost innovation culture and innovators. In Gunasekaran, A. and Sandhu, M. (eds.), *Handbook on Business Information System*, Chapter 15, pp. 359–381. Singapore: World Scientific Publishing Co.

Dhillon, B.S. 2006. *Creativity for Engineers, Series on Industrial and Systems Engineering*, vol. 3, pp. 42–43. Singapore: World Scientific Publishing Co. Pvt. Ltd.

Eklund, T., Paajanen, P., Kantola, J., and Vanharanta, H. 2012. Knowledge creation and learning in organisations: Measuring proactive vision using the co-evolute methodology. *International Journal of Strategic Change Management* 4(2): 190–201.

Ford, M.E. 1992. *Motivating Humans—Goals, Emotions and Personal Agency Beliefs*, p. 248. Newbury Park, CA: Sage Publications, Inc.

Kantola, J. 2005. Ingenious management. Doctoral thesis, Pori, Finland: Tampere University of Technology, Publication 568, p. 183.

Kantola, J., Karwowski, W., and Vanharanta, H. 2010. Managing managerial mosaic: The evolute methodology. In Ordonez de Pablos, P., Lytras, M., Karwowski, W., and Lee, R. (eds.), *Electronic Globalized Business and Sustainable Development Through IT Management: Strategies and Perspectives*, pp. 77–89. Hershey, PA: Business Science Reference.

Kantola, J., Vanharanta, H., and Karwowski, W. 2006. Operators' creative tension and shift performance. *Paperi ja Puu (Paper and Timber)* 88(4): 225–229.

Koskinen, K.U. and Vanharanta, H. 2000. Tacit knowledge as part of engineers' competence. In Vlaovic, B. et al. (eds.). *Proceedings of the Symposium Extra Skills for Young Engineers*, Maribor, Slovenia, October 18–20, 2000, pp. 23–26.

Koskinen, K.U. and Vanharanta, H. 2002. The role of tacit knowledge in innovation processes of small technology companies. *International Journal of Production Economics* 80: 57–64.

Paajanen, P., Piirto, A., Kantola, J., and Vanharanta, H. 2006. FOLIUM—Ontology for organizational knowledge creation. In Callaos, N. et al. (eds.), *WMSCI 2006, Proceedings of the 10th World Multi-Conference on Systemics, Cybernetics and Informatics*, Orlando, FL, July 16–19, 2006, vol. VI, pp. 147–152.

Rauhala, L. 1995. *Tajunnan itsepuolustus (The Self-Defense of Consciousness)*. Helsinki, Finland: Yliopistopaino.

Vanharanta, H. 1995. Hyperknowledge and continuous strategy in executive support systems. *Acta Academiae Aboensis. Ser. B, Mathematica et physica* 55(1): 143s.

Vanharanta, H. 2003. Circles of mind. Identity and diversity in organizations—Building bridges in Europe. In *Programme XIth European Congress on Work and Organizational Psychology*, Lisbon, Portugal, May 14–17, 2003, p. 1s.

Vanharanta, H., Kantola, J., and Karwowski, W. 2005. A paradigm of co-evolutionary management: Creative tension and brain-based company development systems. In *HCI International 2005, 11th International Conference on Human-Computer Interaction*, Las Vegas, NV, July 22–27, 2005, 10pp.

5 Helix
Organizational Commitment and Engagement in Business Organizations

*Jarno Einolander**

ABSTRACT

This chapter describes a novel way to analyze organizational commitment and engagement levels in organizations using an Internet-based evaluation tool. In the application, the respondents estimate the truth value of statements with regard to their respective organization at a given moment in time. The employees also specify how they would like the situation represented by the statements to be in the future. On the basis of the responses, a collective understanding is defined. As a result of the analysis, targets for organization-specific development plans can be planned.

5.1 INTRODUCTION

One of the main sources of competitive advantage for today's organizations is the ability to retain the brightest and most talented employees. In other words, long-term sustained success and growth can be achieved by attracting and retaining the best talent (Heinen and O'Neill 2004). Meyer et al. (1993) recognized organizational commitment as a leading factor impacting the level of achievement in many organizations. Nowadays, as the trend for organizations is to move toward leaner structures, it further emphasizes the importance of committed employees who are engaged in their work. This leads to organizations relying more on fewer employees to do what is needed for the organization to survive and to be successful.

Without a committed and engaged workforce, absenteeism and tardiness can lead to serious problems. Ultimately, undesirable turnover can be critical and extremely costly (e.g., losing productive employees, recruiting, selecting, and training costs, and the potential negative impact on current customer relationships). Therefore, in order to succeed in today's hard competition, organizations must be able to retain their best talent and try to motivate them to raise their engagement levels. This should be one of the most important goals for managers and leaders in today's organizations.

Clearly, identifying employees' level of engagement and commitment is important for business organizations. Evaluating the degree of commitment and engagement, and their types and their sources, facilitates the creation of appropriate policies and

* Tampere University of Technology, Pori, Finland.

practices to foster them in the workforce. Actual "bottom-up" evaluation is important because there is evidence that aggregated employee opinions relate fairly strongly to important business outcomes like performance (Ostroff 1992; Schneider et al. 2003; Vance 2006).

A committed workforce clearly appears to be an advantage for the employer. Employees who are engaged in their work and committed to their organizations give companies significant competitive advantage—they are motivated to work toward their shared goals and objectives, which leads to higher productivity and lower employee turnover. Many studies demonstrate that higher commitment is related to greater levels of satisfaction, motivation, and prosocial behavior, while lower levels are related to a higher intent to quit, a lower turnover rate, and tardiness (Becker and Billings 1993; Cohen 2000). The view that the lack of commitment has a direct relation to employee turnover, and thus a negative impact on the productivity of the organization has been common to all organizational commitment studies for decades (Allen and Meyer 1990). Therefore, management needs tools, that is, management decision support systems with which they can obtain information and knowledge about their employees' levels of engagement and commitment quickly, cost-effectively and systematically, and most importantly in a way that is usable for management.

This chapter gives a description of an Internet-based computer application for evaluating the factors that have an effect on employees' commitment and engagement.

5.2 ORGANIZATIONAL COMMITMENT AND ENGAGEMENT

Organizational commitment means a person's psychological relationship with the organization for which he or she works. Committed employees have been found to contribute to organizational effectiveness if they identify with the organization's goals and values, and are willing to engage in activities that go beyond their immediate role requirements. Mowday et al. (1982) state that commitment is "the relative strength of an individual's identification with and involvement in a particular organization." A common three-dimensional theme is found in most of the definitions of commitment (1) committed employees believe in and accept organizational goals and values, (2) they are willing to devote considerable effort on behalf of their organization, and (3) they are willing to remain with their organization (Mowday and McDade 1979; Steers 1977). Hence, organizational commitment can be described as a psychological state that binds an individual to an organization (Meyer and Herscovitch 2001) and influences individuals to act in ways that are consistent with the interests of the organization (Mowday and McDade 1979; Porter et al. 1974). The advantages of commitment have been found to be increased quality of operations, achievement of objectives, and the continuing development of the organization. Specifically, employees whose commitment is based on affective attachment are more likely to exert effort on behalf of the organization (Allen and Meyer 1990; Mowday et al. 1982).

Engagement is believed to be "one step up" from commitment. Harter et al. (2002), on the other hand, defined employee engagement as "the individual's involvement and satisfaction with as well as enthusiasm for their work." More broadly, Robinson et al. (2004) defined engagement as "a positive attitude held by the employee toward the organization and its values." It has been shown to have an impact on business outcomes, as well as being linked to an increased intention to stay with the organization; as engagement increases, employee turnover decreases (Robinson et al. 2004). Definitions of engagement usually encompass several positive behaviors for organizations. These include behaviors such as the degree to which employees involve themselves in their work, as well as the strength of their commitment to the employer and role. Usually, common themes in engagement definitions reflect concepts like job satisfaction, recognition, pride in the employer, organizational supportiveness, the effort to go the extra mile, and understanding the linkage between one's job and the mission of the organization (Vance 2006). Moreover, engagement is seen to go beyond job satisfaction, referring to an employee's personal state of involvement, contribution, and ownership (Robinson et al. 2004). Many have claimed that employee engagement predicts employee outcomes, organizational success, and financial performance, for example, total shareholder return (Saks 2006). As a result of all of these points, leaders need to have an understanding of how employee commitment develops and is maintained over time (Yousef 2000).

As can be seen, there is a great deal of similarity and overlap between the concepts of organizational commitment and engagement, and their definitions have often been used interchangeably. By definition, the closest type of commitment to engagement is affective commitment, as this type of commitment emphasizes the satisfaction people obtain from their jobs and from their working environment, and the willingness of employees to go beyond their immediate job requirements for the sake of the organization (Allen and Meyer 1990; Meyer and Allen 1991). Affective commitment also goes some way toward capturing the two-way nature of the engaging relationship, as employers are expected to provide a supportive working environment (Robinson et al. 2004). However, commitment does not sufficiently reflect the two-way nature of engagement, and the extent to which engaged employees are expected to have an element of business awareness (Saks 2006). Engagement is not an attitude, but the degree to which an individual is attentive and absorbed in the performance of their occupational roles (Saks 2006).

5.3 HELIX TOOL

For the evaluation of organizational commitment (Einolander and Vanharanta 2011, 2013; Einolander et al. 2011), a computer application named Helix has been used. Helix uses the generic Internet-based platform Evolute (Kantola 2005; Kantola et al. 2006). The application was constructed to contain elements of both organizational commitment and engagement. The ontology behind the application contains features (n = 59) and categories (n = 20). These are assessed with 237 statements. Employees use these statements to evaluate their current reality and their wish or target for the future in their current organization. In this way, employees transfer their own feelings to the system.

The categories that are evaluated are Affective Commitment, Normative Commitment, Continuance Commitment, Intrinsic Motivation, Motivating Potential of Job, Job Satisfaction, Organizational Support, Attributions of HRM Practices, Organizational Dynamics, Person-Role Congruence, Person-Organization Fit

Identification with Organization, Perceptions of Justice and Fairness, Psychological Contracts, Work-Related Investments, Sense of Obligation, Nonwork-Related Investments, and Employment Alternatives and Opportunities.

The statements describing the features and categories were developed based on various studies and models in this field of research. For example, the organizational commitment statements were adapted based on studies by Meyer and Allen (1991), Powell and Meyer (2004), Meyer et al. (2002), and Porter et al. (1974). Statements describing the components of organizational justice were adapted from the Niehoff and Moorman (1993) scale. Job satisfaction and the motivating potential of job measures were developed based on measurement tools devised by Hackman and Oldham (1980) and Weiss et al. (1967). Items relating to the psychological contract were adapted from studies by Raja et al. (2004) and Rousseau (1990, 2000). Role ambiguity and conflict were measured for items based on the Rizzo et al. (1970) scale. About 20% of the answering scales of the statements were reversed, that is, a "positive" value was intentionally changed in order to reduce the response set bias (Hubbard 2010). This was done in order to encourage respondents to read each statement carefully and respond to it accordingly, so as not always to have the answers in a particular direction regardless of the content of the statement.

5.4 CASE

The main goal of the application is to provide a true, comprehensive, "bottom-up" view for management about the state of their employees' commitment and factors affecting their engagement. Identifying levels of commitment is important because only committed employees who are engaged in their work and to their organization are those who really enable an organization to grow and flourish by giving a competitive advantage, including higher productivity and lower employee turnover. It has been argued that, by using a professionally developed assessment tool, on average management will be more effective at making employment-related decisions than by just simply observing the employees (Rivkin 2000). Use of the application provides important information for management about their employees collectively and gives guidance for prioritizing potential development activities by comparing the current reality to the desired target state. The results of the evaluation can be used as a basis for open discussion when creating development activities for the organization, as well as for educating and learning about these concepts.

Section 5.5 illustrates some sample results of the analysis. The data for this sample were gathered during the spring of 2014 from a Finnish energy company operating in the electricity and district heating market. The case data contain results from 10 senior salaried staff in various expert positions.

5.5 RESULTS

Figure 5.1 represents an example of the category level results of the preliminary tests. The upper dark gray colored bars represent the group's collective perception of the current reality (perceived current state), and the lower light gray colored bars represent their vision for the future, and the difference is called the collective proactive

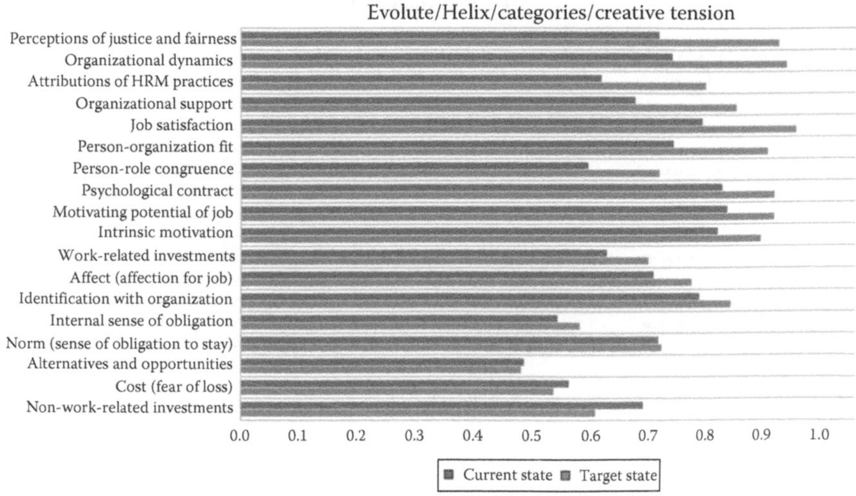

FIGURE 5.1 Example of category-level results.

vision or creative tension. The results have been sorted based on the highest proactive vision, that is, the greatest collective feeling of tension between the current and envisioned future state.

When looking at Figure 5.1, for example, it seems that the test group feels collectively that there is most room for improvement or change in "Perceptions of Justice and Fairness." "Perceptions of Justice and Fairness" contains features describing how employees perceive that they have been treated fairly in their jobs. If employees perceive that they are treated fairly, or they perceive the process by which outcome allocation decisions are made as being fair, they are likely to reciprocate by doing actions beneficial to the organization that go beyond the normal performance requirements of their jobs (Niehoff and Moorman 1993). The category with the second highest creative tension is "organizational dynamics," which includes features describing open communication, information sharing, and participative decision making in the organization between employees and employers (Choudhury 2011) and strategic alignment, describing a familiarity with the organization's strategy and how employees' actions contribute to its fulfillment, and whether they put their effort in the right direction. In this case, this category should be looked at on the feature level to see where the most tension lies.

Figure 5.2 presents the averages and standard deviations of the same category level results as in Figure 5.1. The dark gray bars represent the current state results and their standard deviation in the research group. Likewise, the light gray bars represent the range of category level results and their standard deviation in the future target state of the research group. The lines represent the averages of these results. It is important to consider the distribution of responses alongside other measures. Looking at standard deviations gives an indication of how far the individual responses vary or "deviate" from the mean; in other words, it shows how spread out the responses are and whether there is a high degree of unity between the respondents.

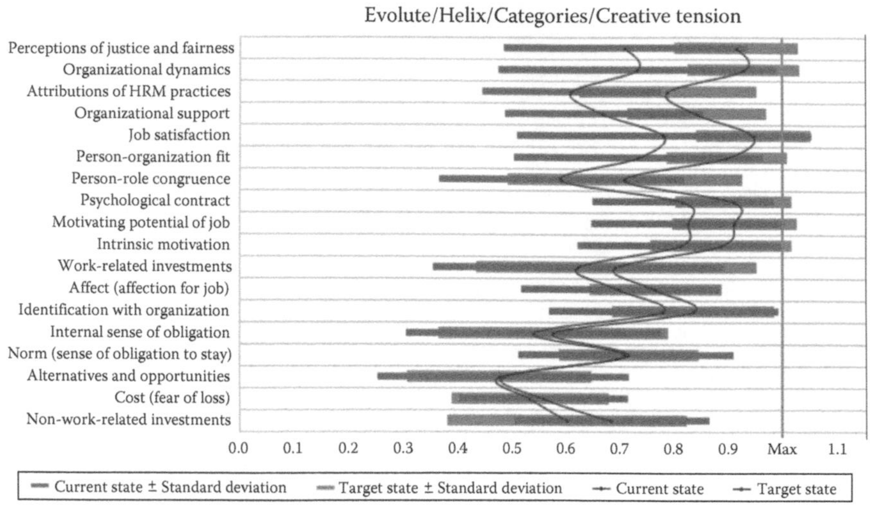

FIGURE 5.2 Average and standard deviation of category level results.

In Figure 5.2, when considering the same categories with the highest creative tension as in Figure 5.1, it becomes evident that even though there is high creative tension, the responses also have a very high deviation between them especially in the current state.

In order to delve into what lies behind the category level results and to pinpoint the most appropriate targets for possible development activities, we must look at the feature level results. Figure 5.3 shows the 10 feature-level results with the highest proactive vision results in the sample (n = 10).

Figure 5.3 portrays 10 features with the highest proactive vision in the sample. For example, it can be seen that the highest tension between the current state and target state is in satisfaction with the compensation received and information sharing. Compensation satisfaction is a difficult feature because, if asked about directly, it will probably be regarded as inadequate. However, if not asked about directly, it is

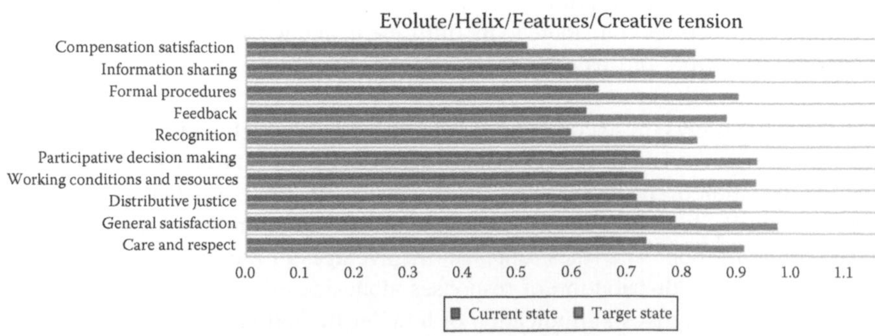

FIGURE 5.3 Example of variable level results of the third study.

likely it will be not stressed as much. In the top 10 highest creative tension features, there are also features concerning justice and fairness, decision-making practices, information sharing, and feelings of respect shown to the employees.

These features are those that the respondents feel their employer should direct its development practices toward and improve. This may lead to increased overall engagement in a specific organizational setting. If you wish to examine what affects these feature level results the most, they can be investigated more deeply by looking at the individual statements they are comprised of, and the responses. However, the standard deviation and distribution of responses should be examined in order to understand the group of respondents more deeply. This is important when planning concrete HR development actions.

5.6 DISCUSSION

There are a wide variety of reasons why organizational commitment is important and why organizations want their key employees to be highly committed. Studies have demonstrated that higher commitment is related to factors such as greater levels of satisfaction, motivation, and prosocial behavior, while lower levels are related to a higher intent to quit, a higher turnover rate, and tardiness. In addition, high levels of organizational commitment have been associated with greater work attendance, extra role behaviors, and reduced levels of absenteeism. Mathieu and Zajac (1990) concluded organizational commitment to be a useful criterion for various organizational interventions designed to improve employees' attitudes and behaviors. At a minimum, they suggest that it should be used to influence the employees' socialization processes, participation, ownership in the company, and reactions to job enrichment.

It is hard to measure concepts like commitment or engagement because it requires the assessment of the complex feelings and emotions of employees. The factors affecting each person's feelings and emotions vary from individual to individual; for example, based on their personality, situation in life, values, or goals. When assessing an organization or employee group within one organization, for example, there will probably be a different set of factors affecting engagement and commitment. In this chapter, we described our computer application that was designed to tackle this problem.

We developed this tool to make the factors affecting commitment clearer for management. The visual nature of this tool will enable management to gain a better understanding of this concept and quickly see the degree of the factors affecting it for a certain specific group.

Understanding the factors or personal characteristics that enhance commitment and engagement has practical value for organizations. Management should discover how their employees feel and find out where they see most flaws in their work environment. Management should then try to make plans to correct these drawbacks and to show employees that they are valued, while keeping the organization's best interests in mind. This should lead to greater motivation and employees becoming more engaged and committed.

This information will also help management to compare results of various evaluated groups and plan-specific actions. We believe that, in addition to scale validity and reliability, the measurement method and the results it provides must be useful with regard to the organization's goals and objectives, and it must be possible to administer it cost-effectively and rapidly.

It is easy for management to collect employee opinions widely and effectively using a personnel survey. However, such tools can only give rough guidelines for the matter in hand because of the differences in people, culture, and the constantly changing environment. Nevertheless, they provide management with considerable evidence about employees' feelings and information for decision support. Therefore, managers and leaders should actively use this valuable information to improve decisions, and not ignore its potential.

REFERENCES

Allen, N.J. and Meyer, J.P. 1990. The measurement and antecedents of affective, continuance and normative commitment to the organization. *Journal of Occupational Psychology* 63(1): 1–18.

Becker, T.E. and Billings, R.S. 1993. Profiles of commitment: An empirical test. *Journal of Organizational Behavior* 14(2): 177–190.

Choudhury, G. 2011. The dynamics of organizational climate: An exploration. *Management Insight* 7(2): 111–116.

Cohen, A. 2000. The relationship between commitment forms and work outcomes: A comparison of three models. *Human Relations* 53(3): 387–417.

Einolander, J. and Vanharanta, H. 2011. Basics of ontology-based organizational commitment evaluation. *Global Partnership Management* 48–55.

Einolander, J. and Vanharanta, H. 2013. Organizational commitment among purchasing and supply chain personnel. *International Journal of Digital Information and Wireless Communications (IJDIWC)* 3(4): 43–50.

Einolander, J., Vanharanta, H., and Kantola, J. 2011. Managing commitment through ontology-based evaluation: Basics of new decision support system. In *Proceedings, 2011 World Congress on Engineering and Technology, CET 2011*, Shanghai, China, October 28–November 2, 2011, pp. 167–171.

Hackman, J.R. and Oldham, G.R. 1980. *Work Redesign*, 330pp. Reading, MA: Addison-Wesley.

Harter, J.K., Schmidt, F.L., and Hayes, T.L. 2002. Business-unit-level relationship between employee satisfaction, employee engagement, and business outcomes: A meta-analysis. *Journal of Applied Psychology* 87(2): 268.

Heinen, J.S. and O'Neill, C. 2004. Managing talent to maximize performance. *Employment Relations Today* 31(2): 67–82.

Hubbard, D.W. 2010. *How to Measure Anything: Finding the Value of Intangibles in Business*, 2nd edn. Hoboken, NJ: John Wiley & Sons, Inc.

Kantola, J. 2005. Ingenious management. Tampere, Finland: Tampere University of Technology. Publication, 568.

Kantola, J., Vanharanta, H., and Karwowski, W. 2006. The Evolute system: A co-evolutionary human resource development methodology. *International Encyclopedia of Human Factors and Ergonomics* 3: 2902–2908.

Mathieu, J.E. and Zajac, D.M. 1990. A review and meta-analysis of the antecedents, correlates, and consequences of organizational commitment. *Psychological Bulletin* 108(2): 171.

Meyer, J.P. and Allen, N.J. 1991. A three-component conceptualization of organizational commitment. *Human Resource Management Review* 1(1): 61–89.

Meyer, J.P., Allen, N.J., and Smith, C.A. 1993. Commitment to organizations and occupations: Extension and test of a three-component conceptualization. *Journal of Applied Psychology* 78(4): 538–551.

Meyer, J.P. and Herscovitch, L. 2001. Commitment in the workplace: Toward a general model. *Human Resource Management Review* 11(3): 299–326.

Meyer, J.P., Stanley, D.J., Herscovitch, L., and Topolnytsky, L. 2002. Affective, continuance, and normative commitment to the organization: A meta-analysis of antecedents, correlates, and consequences. *Journal of Vocational Behavior* 61(1): 20–52.

Mowday, R.T. and McDade, T.W. 1979. Linking behavioral and attitudinal commitment: A longitudinal analysis of job choice and job attitudes. *Academy of Management Proceedings* 1979(1): 84–88.

Mowday, R.T., Porter, L.W., and Steers, R.M. 1982. *Employee-Organization Linkages: The Psychology of Commitment, Absenteeism, and Turnover*, vol. 153. New York: Academic Press.

Niehoff, B.P. and Moorman, R.H. 1993. Justice as a mediator of the relationship between methods of monitoring and organizational citizenship behavior. *Academy of Management Journal* 36(3): 527–556.

Ostroff, C. 1992. The relationship between satisfaction, attitudes, and performance: An organizational level analysis. *Journal of Applied Psychology* 77(6): 963.

Porter, L.W., Steers, R.M., Mowday, R.T., and Boulian, P.V. 1974. Organizational commitment, job satisfaction, and turnover among psychiatric technicians. *Journal of Applied Psychology* 59(5): 603.

Powell, D.M. and Meyer, J.P. 2004. Side-bet theory and the three-component model of organizational commitment. *Journal of Vocational Behavior* 65(1): 157–177.

Raja, U., Johns, G., and Ntalianis, F. 2004. The impact of personality on psychological contracts. *Academy of Management Journal* 47(3): 350–367.

Rivkin, D. (2000). *Testing and Assessment: An Employer's Guide to Good Practices.* Washington, DC: U.S. Department of Labor, Employment and Training Administration, Office of Policy and Research.

Rizzo, J.R., House, R.J., and Lirtzman, S.I. 1970. Role conflict and ambiguity in complex organizations. *Administrative Science Quarterly* 15: 150–163.

Robinson, D., Perryman, S., and Hayday, S. 2004. The drivers of employee engagement. Brighton, U.K.: Report-Institute for Employment Studies.

Rousseau, D.M. 1990. New hire perceptions of their own and their employer's obligations: A study of psychological contracts. *Journal of Organizational Behavior* 11(5): 389–400.

Rousseau, D.M. 2000. Psychological contract inventory: Technical report. Boston, MA: British Library.

Saks, A.M. 2006. Antecedents and consequences of employee engagement. *Journal of Managerial Psychology* 21(7): 600–619.

Schneider, B., Hanges, P.J., Smith, D.B., and Salvaggio, A.N. 2003. Which comes first: Employee attitudes or organizational financial and market performance? *Journal of Applied Psychology* 88(5): 836.

Steers, R.M. 1977. Antecedents and outcomes of organizational commitment. *Administrative Science Quarterly* 22: 46–56.

Vance, R.J. 2006. *Employee Engagement and Commitment.* Alexandria, VA: SHRM Foundation.

Weiss, D.J., Dawis, R.V., England, G.W., and Lofquist, L.H. 1967. *Manual for the Minnesota Satisfaction Questionnaire*, vol. 22, *Minnesota Studies in Vocational Rehabilitation.* Minneapolis, MN: University of Minnesota.

Yousef, D.A. 2000. Organizational commitment: A mediator of the relationships of leadership behavior with job satisfaction and performance in a non-western country. *Journal of Managerial Psychology* 15(1): 6–24.

6 Kappa
Future Challenges for Sales Function

*Kari Ingman**

ABSTRACT

Culture, including company culture, is like a person's character: highly permanent in nature and hard, if not impossible, to change. Company culture is difficult to describe explicitly in a few sentences because it is so deep in an organization's way of thinking and performing daily routines. It is very important for sales managers to comprehend culture's influence on the whole company, albeit challenging and time consuming. Subcultures, which inevitably emerge when organizations mature, are needed for renewal but might also cause tearing contradictions, which paralyze the whole company.

6.1 INTRODUCTION

Sales work is facing great challenges as new e-commerce-based sales channels are expanding and becoming more and more important. As a direct consequence, the number of salespeople is decreasing dramatically. Research company Gartner estimates the decrease in the United States to be from 18 billion salespeople to 4 billion by 2020. Even though the magnitude is not that massive in Finland, the direction of sales work development is clear: the quantity of salespeople conducting personal sales work is decreasing.

Even with fewer salespeople, the value of personal sales work is not clearly reducing. The customer still needs salespersons' expertise and support when purchasing a product, service, or solution. The challenge is that the customer has proceeded far in purchasing process before contacting the salesperson, because customers gather data for purchasing from for example, the Internet. Hence, the salesperson has fewer opportunities to influence the purchase process than before—especially in the early stages of the process. Rapid changes in market situation, more harsh competition, and advent of new, cost-effective sales channels to the marketplace affect sales work and thereby also the management of the sales (Majamäki 2012).

Another major challenge for a successful sales manager is to be able to manage and lead a sales force in the global marketplace. This is vital even if a company is not operating in export markets, because hardly any market is not impacted by global forces. Hence, also at home, market companies must be prepared to compete against global competitors. It is a fact that globalization will continue at a rapid pace and if a firm is unable—or unwilling—to satisfy global customers, competitors will (Ford et al. 2003).

* Tampere University of Technology, Pori, Finland.

6.2 IMPORTANCE OF CULTURE FOR SALES MANAGEMENT

The challenge for sales management is to support salespersons in their daily sales work and also to foreseen future requirements of development for an individual salesperson. Also, customers demand more added value, better product benefits, and solutions that even exceed expectations from salespersons. To meet these expectations, sales management must ensure that salespersons have understanding of customer business environment and support from own organization. Eventually, sales managers' chief responsibility is to ensure that sales force actions produce satisfied customers in a long sustaining base (Majamäki 2012).

Changes in the business environment and marketplace require changes in an organizations' way of thinking and performing—from a strategic level to daily operations. First, the emerging challenges must be identified, and second reacted on. Fundamental changes in business environments demand decisive changes in an organization, but how is one to ensure that completed actions will have a positive result? Or how to be sure that changes made would last and become an intrinsic part of day-to-day work? The answer is to understand company culture, subcultures such as sales culture and especially the processes to change them. To get a picture of one's own sales organization's culture, the sales management clearly needs a tool to evaluate the sales culture status.

Kappa tool, together with a workshop, is designed to give an insight into culture, current level of an organization's culture, and perceive areas for improvement, thus allowing development resources to be allocated accordingly.

6.3 SALES CULTURE ONTOLOGY AND KAPPA TOOL

The purpose of this study is not to go deep into a perception of culture, emphasizing the following quotation: "Central elements of any culture will be the assumptions the members of the organization share about their identity and ultimate mission or functions." To understand sales culture in a company comprising multiple functions (production, R&D, service, sales, marketing), an understanding of subculture is essential. In addition, for renewal, which is in the long term the only way to survive, an understanding of the subculture is vital. According to Hofstede (1982) outstandingly successful organizations usually have strong and unique subcultures.

Organizational learning, development, and planned change cannot be understood without considering culture as a primary source of resistance to change. If it is understood that most organizational changes involve changes, especially at the subcultural level, the challenge lies in conceptualizing a culture of innovation in which learning, adaptation, innovation, and perpetual change are the stable elements (Schein 2004).

The creation of subcultures follows a certain pattern; if the organization is successful and expands, it inevitably creates smaller units that begin the process of culture formation on their own with their own leaders.

The major bases on which such differentiation occurs are as follows (Schein 2004):

1. Functional/occupational differentiation
2. Geographical differentiation
3. Differentiation by product, market, or technology

4. Divisionalization
5. Differentiation by hierarchical level

As an example of occupational differentiation that has caused the emergence of two subcultures (in this case, sales and marketing cultures) and conflicts between them which weakens the culture of the parent company: One of the reasons that marketing and sales departments often develop communication problems with each other is because salespeople develop part of their culture from their constant interaction with the customer, whereas the marketing group is generally more immersed in the headquarters' culture and its technical subculture. Often, marketing sees itself as creating the strategies and tactics that sales must then implement, leading to potential status conflicts. The important point to recognize is that the difficulty often encountered between these functions can be seen to result from genuine subcultural differences that are predictable and can be analyzed (Schein 2004).

Company culture is defined by mission, values, and strategy. Sales, which is a subculture of a company's culture, has its own strategy but must naturally be in concordance with the company strategy. If consensus fails to develop, and strong subcultures form around different assumptions, the organization will find itself in serious conflict that might undermine its ability to cope with the external environment. Sales culture consists of five main concepts: sales culture support, sales force, sales system, sales process, and sales culture assumptions.

Sales culture support has three subconcepts. The first contains suitable elements of company strategy, vision, and values. Second is the ethics of sales. For sales, ethics has probably the most important role in the company, because sales processes vary a lot depending on the customers and market areas. The opportunity to act in an unethical way is the biggest hurdle during sales and marketing activities: for example, selling something unsuitable or considering own personal benefit. The third and most important subconcept is leadership, which is to build, change, and strengthen the culture. The leadership subconcept can further be divided into leadership, management, and time management, but to keep this ontology lean enough, the term leadership presents all three within this study.

Elements of the *sales force* concept establish paths through which a single sales activity takes form. These paths can be seen in Figure 6.1 as dotted lines. Some of the elements can be regarded as tools of leadership, and those are affected by leaders, but some of them will be developed over a longer time period and cannot be controlled directly. Elements along a path constitute a logical chain that is behind a single action: for example, training and development improve the skills of a salesperson and thus improving understanding of selected actions. Following the elements on a path, leaders and managers can focus the sequential development efforts on correct issues.

Sales system defines the objects for sales activities and in which market areas these activities should be performed. The general sales strategy is set in the sales system subconcept, defining a company's strategic position in the marketplace, and when leaders/top managers communicate this strategy with the whole organization, the sales system is the detailed projection of the strategy from a sales point of view. The sales system also includes prescription of functions and resources outside sales

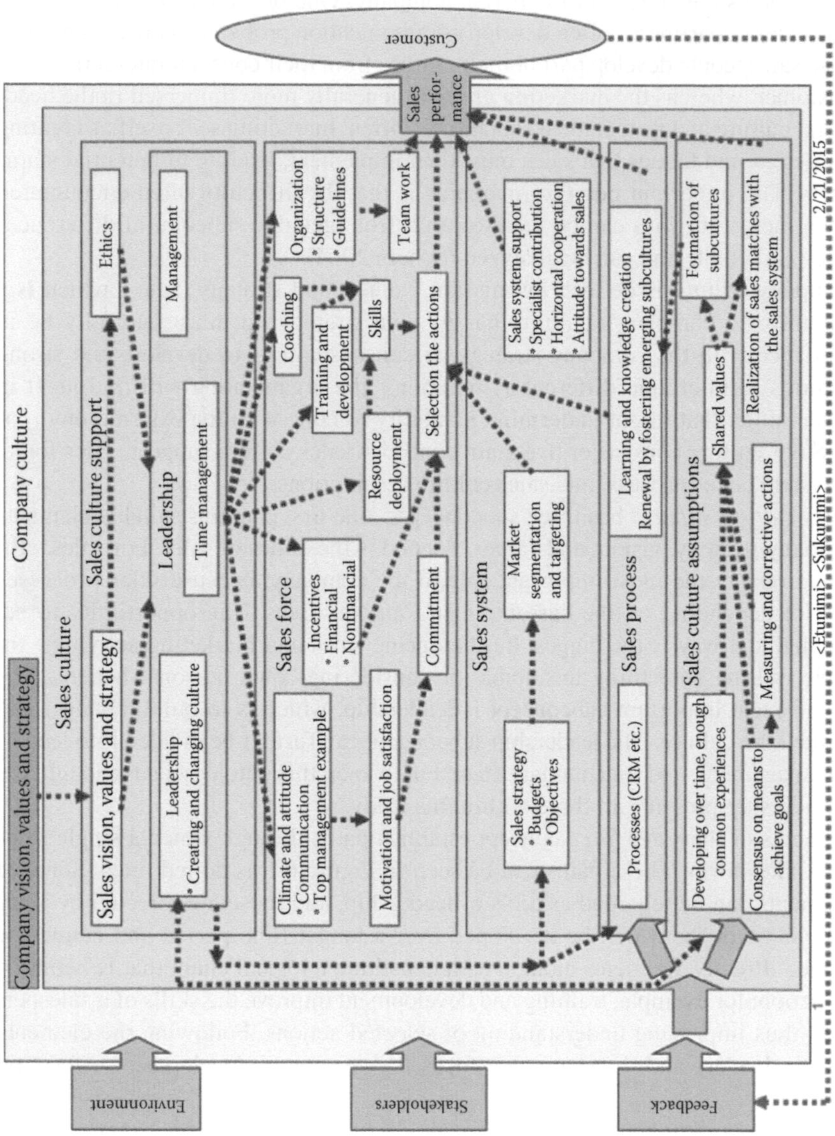

FIGURE 6.1 The sales culture ontology.

organization, which are needed to support sales processes, for example, the more technical products company is selling, the more technical expertise and support from other functions are needed. Essential for an organization's success is the attitude of how sales is regarded and how well sales is integrated to the other functions. If sales remains as an isolated function, it is easy to understand that a company loses many opportunities to improve customer service; in customer meetings, it might be hard to convince a customer without experts, and on the other hand, if salespersons are not listened to, valuable information from the marketplace might not reach the technical area at all.

Sales process concept comprises the theoretical context of the sales procedure. Described here are all of the processes that are needed in the company to realize sales efforts. The quantity of processes varies depending on the company's business. The simplest sales process is for agencies, which have only few major customers on the domestic market. The most complicated processes are for multinational companies, which have their own sales offices and agencies serving customers globally. A vital part of the sales process is continuous learning and knowledge creation. If training, which is included in the sales force concept, is more for individual salespersons to improve their sales competencies, learning, and knowledge creation includes the processing of sales-related innovations and processes, where tacit knowledge is converted to explicit knowledge. Continuous learning and knowledge creation shall be a controlled process to ensure proper results.

Sales culture assumptions reflects the underlying assumptions that guide salespeople's professional decisions and daily actions. Most of these assumptions are not conscious but operate behind decisions. These assumptions develop over time, and it could be said that if any assumptions cannot be found, there is no culture or the culture is weak. Furthermore, in case of young companies, it is natural that elements of culture have not yet evolved. Leaders have a very strong effect on the progression of the sales culture, especially in the early stages of organizational development. Typically, an organization follows the example of the leader (in many cases the leader is also the founder): what he/she pays attention to, how he/she reacts to certain situations, etc. By repeating certain patterns of behavior, leaders start to build the underlying assumptions for the organization. When an organization as a whole can agree about the objectives and applicable methods to reach the objectives, the culture starts to form and strengthen. One sign of a more advanced culture is the consensus over measuring the results, and especially a consensus over the actions needed if objectives are not reached. Leaders can develop and change the culture, but the culture exists among the individuals of the organization. Therefore, culture interprets how well the sales strategy is realized in daily operations, sales systems, and sales processes. Culture is about activities that are performed daily. The difference between defined and actual behavior reflects the level of culture. It is essential that the group shares values, which are important for them and from the company's point of view, and those values should be in harmony with the company's values. When an organization grows larger, it is natural that several groups in it would develop different value bases. This is acceptable as long as the most important values are the same for all, and a single group's values do not contradict the company's.

Sales culture can be seen through customers' eyes as sales performance by a sales force. A theoretical "best way" for the sales process exists, but how the underlying assumptions affect this in practice, that is, whether or not the written method is followed in practice, is the core of sales culture. It is easy to put on paper an ideal way to handle a sales case and other customer management actions, but in reality, what is crucial is how well each salesperson operates according to the explicitly defined way.

As shown in Figure 6.1, there are several elements influencing sales performance. This influence is shown as dotted lines ending in sales performance. Based on how a customer finds the sales performance, positive or negative feedback is given and this feedback is handled mainly in the sales process concept. Customer feedback also affects the sales culture assumptions because there is a two-way connection from the sales culture assumptions to leadership and back. The received customer feedback is also handled by leaders because they can change the culture in the most powerful way. Feedback-based changes can often be tracked to leadership.

6.4 CASE

The technology company that was studied in this case had a strong expansion of its sales force outside domestic markets. Their sales personnel sell the same products in various market areas where the cultural background of customers varies significantly. Single salesmen have remote locations on different continents without a close connection to the company's technical staff, and hence, an understanding of multicultural and cooperation is needed in the headquarters. Cultural understanding is also required to interpret customers' feedback via local salespeople from various territories.

The objective in this case was first to define the culture in an industrial environment and second to build an ontology of sales culture and its relation to organizational culture in order to understand underlying processes within a sales culture. The third objective was to use the new ontology to study and analyze the export sales force in the company to find out which areas of the sales process need to be developed. The fourth objective was to see how well the results match with the literature defining the most important factors (organizational competencies) to create a strong sales culture. These factors are embedded in Kappa statements. Finally, a prioritized list of actions was suggested for strategic decision making based on individuals' collective inner perceptions.

From a total or 19 inquiries, a completed evaluation was received from 14 participants, so the answering ratio was 74%. After all the employees completed the inquiry, the results were summarized so that no individual's answer could be seen. The results thus show the collective status in a whole sales organization. All results are shown in graphical format.

6.5 RESULTS

In Figure 6.2, the results are sorted top-down according to the biggest proactive vision (Vision–Current). The bigger the proactive vision, the more employees want to improve the concept. It can be anticipated that development efforts focusing on

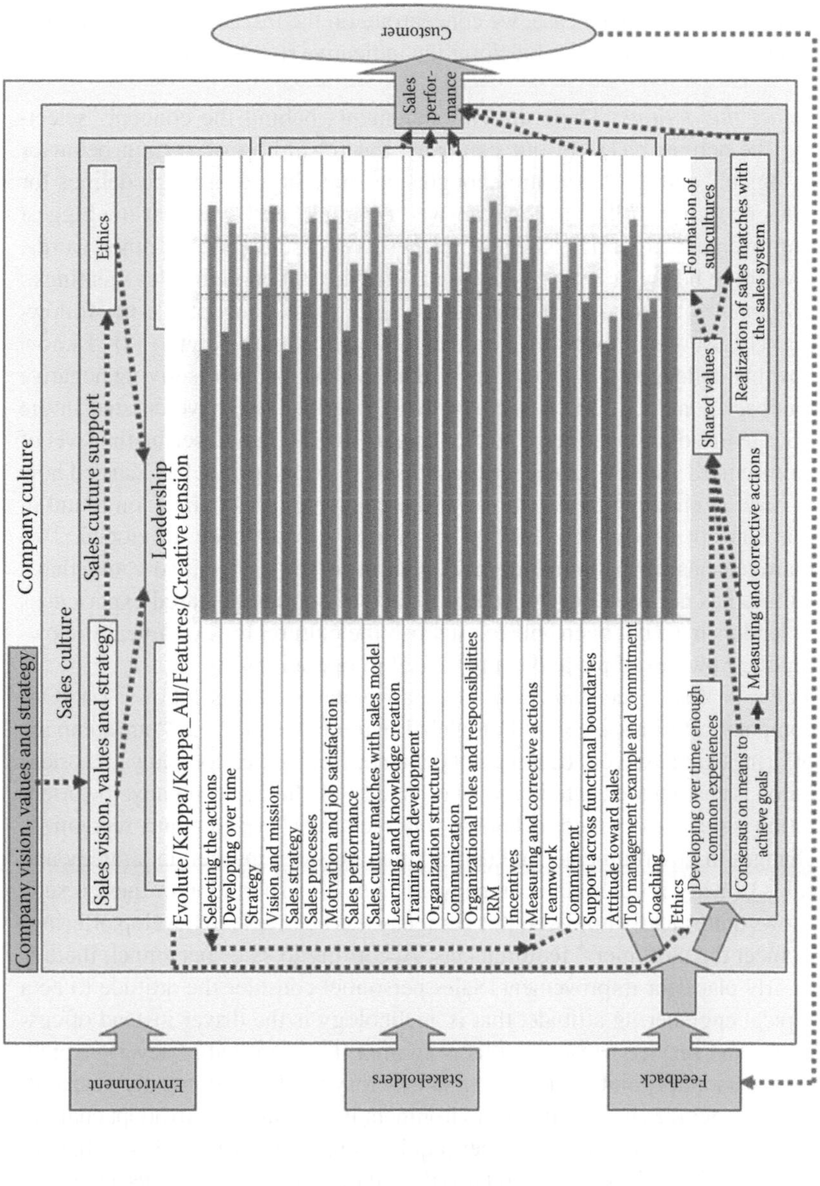

FIGURE 6.2 Concepts in the sales culture ontology sorted according to the biggest proactive vision.

the concepts having the biggest proactive vision would give the best results due to a high internal demand of employees. In this case, the biggest proactive vision was for "selecting the actions," the second was "development over time," and the third was "strategy." At the other end, "ethics" has practically no proactive vision, meaning that employees do not see any need/place for improvements in ethics-related issues. Based on this, it is easy to understand that efforts to improve ethics in the company will not be successful. In this Kappa case, we concentrate on the three concepts with the biggest proactive vision, and look a bit closer the indicative statements behind the results.

Selecting the actions: There are two statements behind the concept "selecting the actions": (1) "In our company, the rewarding of certain behavior is logical" and (2) "there are enough sales meetings to give guidelines for sales operations." The average answer by employees was that the biggest improvement potential is found in these two indicators; in other words, rewarding behavior is not logical and there are not enough sales meetings.

Development over time: The statements behind this concept are the following: (1) "New members in sales team adapts the team habits," (2) "I know how my colleague team members will act when customer is giving negative feedback," and (3) "when one have work overload, we have created in the sales team solutions to solve it." Statements in this class describe the level of culture in the most original form, that is, how well a group has learned how to react in challenging situations. If there is a big proactive vision found in this concept, it is quite clear that the culture has not been developed. Two main reasons can be found for this: either the organization is so young that a culture has not been developed due to lack of common, shared experiences or leaders have not been able to establish the culture, for example, by showing their own example and being logical in operations.

Strategy: In this concept, there are indicative statements as follows: (1) "Our company offerings are based on global market demand," (2) "our company offerings are based on local market demand," (3) "in our company, the orientation is customer oriented/product oriented," (4) "in our company, the orientation is service oriented/technology oriented," (5) "our company recognizes changes in environment," and (6) "relation between marketing strategy and sales strategy is defined in company strategy." Interpretation of these results shows quite clearly how well this company has been able to develop offerings to meet the customers' requirements. According to sales personnel, there is clearly place for improvement. Sales personnel consider the attitude to be a typical engineering attitude, that is, technology is the driver instead of customer and service. However, this is an area that is probably never—from a sales personnel point of view—at a satisfactory level. The company's capability to recognize changes in the environment is vital in order to adapt changes to the strategy and meet customer requirements in the future. If no changes are recognized, the company retains the old methods and offerings and inevitably disaster follows. The last statement refers to the roles of sales and marketing, and especially to the relationship between them, which is quite often poorly defined. In the literature, it is often seen that in an organization, the

> sales department, even when it has a large number of personnel, is located under marketing. However, a distinction in roles between these two organizations shall be clear in order to avoid overlapping by performing the same tasks simultaneously or leaving some tasks completely without attention.

To find out the level of the case company's sales culture, one option is to compare the results to aspects that can be found in the literature regarding the evolution of a strong culture. The literature gives the following elements as the main components when creating a strong culture: (1) leaders' role: they show their own example and operate logically (Oedewald and Reiman 2003; Smith and Rutigliano 2001), (2) organization acts as defined in process descriptions (Nivaro 2008; Oedewald and Reiman 2003), (3) personnel is committed to work and company (Oedewald and Reiman 2003), (4) culture develops over time, that is, groups have had enough common experiences (Schein 2004), and (5) sales strategy is clear to everyone (Oedewald and Reiman 2003; Schoepf 2007). Elements found in the literature do not exactly match with the concepts in the SCO, but there are still clear similarities.

Recommended actions: The culture is a pattern of shared basic assumptions that the group learned as it solved its problems of external adaptation and internal integration that have worked well enough to be considered valid and, therefore, to be taught to new members as the correct way to perceive, think, and feel in relation to those problems (Schein 2004). If top management is not supporting this evolution taking place within a group by operating itself systematically, it weakens the culture.

Consensus about illogical and even unfair rewarding is obvious among the sales force. The recommended action for the management is to clarify what has been the basis for rewarding, analyzing it, and making new guidelines for rewarding. The final step is to inform the whole sales force about these guidelines and to commit to them by top management. The literature supports this; the leaders' role is very important. Rewarding here must be seen quite widely, as it is not only incentives but also promotions, getting certain sales territories, getting customers or customer groups, being allowed to operate with larger authorizations or public rewarding; for example, in a big sales case, if a top management take-over of the case occurs, this will lead to an unsatisfactory situation from a salesperson's point of view, as a top management role should be more supportive in order to improve a salesperson's knowledge.

Culture development, which takes place in groups, needs forums, and face-to-face meetings. For this purpose, sales meetings are constructive and natural events. Disseminating information can be done during these meetings. It would be easy to propose more sales meetings for the whole sales force, but limitations quickly build up due to expenses and time usage. The way to improve this issue is to develop the quality of the meetings, use modern negotiation media (net meetings) instead of travelling and increasing the number of local meetings. Support from the literature can be found here, too: a culture develops over time, that is, when groups have had enough common experiences.

Subcultures among sales forces are another issue that should be controlled and supported to a certain extent. An emerging strong subculture might cause turbulence and bring down the performance of the total sales force. During sales meetings, the sales management has a good opportunity to take an overview of the subculture status within a sales force and—if needed—take corrective actions. On the other hand, management should understand why new subcultures are arising; there might be something valuable behind this, which could be a basis for new businesses, new processes, or a new approach to existing issues.

6.6 DISCUSSION

In this study, a clinical research model was utilized, as outlined by Schein (2004) as the most appropriate methodology for cultural deciphering. The critical distinguishing feature of the clinical research model is that the data come voluntarily from the members of the organization because either they initiated the process and have something to gain by revealing themselves to the clinician, consultant, or researcher, or if the consultant initiated the project, they feel that they have something to gain from cooperating with the consultant. This method is more powerful than other methods because if the researcher/consultant is helping the organization, he/she is therefore licensed to ask all kinds of questions that can lead directly into cultural analysis and thereby allow the development of a research focus too. Both the consultant and the "client," therefore, become fully involved in the problem-solving process and the search for relevant data becomes a joint responsibility. To complete the sales culture survey and conduct actual changes, it is necessary that the consultant is involved in implementing the corrective actions. When leaders are committed to accomplish these corrective actions, there is a realistic chance of changing the culture.

In general, strong divisional subcultures will not be a problem to the parent organization unless the parent wants to implement certain common practices and management processes. From a leadership point of view, there is a clear challenge for leaders: the leader must be sensitive to different subcultures and must develop the skills of working across cultural boundaries. Building an effective organization is ultimately a matter of meshing the different subcultures by encouraging the evolution of common goals, common language, and common procedures for solving problems (Schein 2004).

REFERENCES

Ford, J.B., Honeycutt, E., and Simintiras, A. 2003. *Sales Management: A Global Perspective*. London, U.K.: Routledge.

Hofstede, G.H. 1982. *Cultures Consequences, International Differences in Work-Related Values*. Beverly Hills, CA: Sage Publications.

Majamäki, M. Miten käy myynnin johtamisen murroksessa. Kauppalehti 8.10.2012.

Nivaro, H. 2008. Improvisointi kuriin, myynnin johtaja. *Fakta: talous ja tekniikka tänään* 12: 44–45.

Oedewald, P. and Reiman, T. 2003. Core task modelling in cultural assessment: A case study in nuclear power plant maintenance. *Cognition, Technology & Work* 5: 283–293.

Schein, E.H. 2004. *Organizational Culture and Leadership*, 3rd edn. San Francisco, CA: Jossey-Bass.

Schoepf, D. 2007. What's a sales culture? http://ezinearticles.com/?Whats-A-Sales-Culture?&id=401954. Accessed on January 2, 2007.

Smith, B. and Rutigliano, T. 2001. Creating a successful sales culture. http://www.gallup.com/businessjournal/328/Creating-Successful-Sales-Culture.aspx. Accessed on November 26, 2001.

7 Chronos and Kairos *Understanding and Managing Time*

*Tero Reunanen**

ABSTRACT

Time is the most important resource in a modern business environment. Every other resource can be added or reduced, but time is imperative. Resources such as personnel, capital, or facilities are also needed, but their usage is always time dependent. People's productivity, and hence organizational performance, is heavily related to time usage. On the one hand, some people can be very productive in a very short time period, even when work is not very efficient, and on the other, a person who is very efficient may even damage the organization. Therefore, organizations and especially leaders should focus on where they use their time.

Time is not an easy concept to handle or understand. Time has many different faces toward people. The challenge is that chronological time, Western business culture time, is not suitable when human relations are handled. Every individual experiences time differently, and different situations constantly change the experience. Therefore, it is hard to have schedules to match or plans to actualize within a scheduled time for different individuals. Business is normally done and agreed in terms of chronological time, and the work is divided for leaders into smaller and smaller portions. Consequently, it is crucial for organizations to understand how its members experience their time and how time can be taken into account. Before it is possible to manage one's own time usage, personal time orientation, biases toward time, and situation have to be understood consciously. This chapter handles the Chronos and Kairos application, which is developed to help leaders in their way toward time management.

7.1 INTRODUCTION

Generally, time is not an easy concept to master. Humans have learned different ways to handle time. It is possible to measure time's duration, speed, and numerical order with clocks (Sorli 2002), but this is far from really understanding time. There are different points of views in terms of understanding time. Newtonian physics' point of view toward time is that it is independent of other physical phenomena and is absolute. On the other hand, in general relations theory, it is stated as being a fourth dimension of space, and the changes in it are irreversible. For instance, Sorli states

* Turku University of Applied Sciences, Finland.

that "physical time exists only as a stream of change" (Sorli and Sorli 2004). There are also different ways to connect time to something that is more easy to realize. Boroditsky (2000) proposes that time can be understood as a spatiotemporal metaphor, and this causes relations between space and time. However, she also submits that there is no evidence that these metaphors are necessary when thinking of time. Despite the various definitions or metaphors, what is meaningful for managers is that time is a unique resource that cannot be stored, and time is perishable and irreplaceable and has no substitute. Demand does not affect it, and it has no price or marginal utility curve. One thing above all, when talking about time in modern working life, is that it is always short of supply, that is, we are always lacking it (Drucker 2005, p. 226; Turnbull 2004). Keeping in mind these factors, it is crucial for managers to manage their time and use it wisely.

One of the most important skills for leaders is the skill of managing oneself. Drucker (2005) starts his article in *Harvard Business Review* with the sentence: "Success in knowledge economy comes to those who know themselves—their strengths, their values and how they best perform." In the article, Drucker handles personal management skills and emphasizes consciousness about oneself. Drucker has also observed that "effective executives do not start with their tasks, they start with their time." This way, he underlines that time is a limiting factor (Drucker 1967, pp. 25–26). Because of the nature of knowledge and managerial work, the difference between either wasting time or using it wisely is a matter of effectiveness and results (Drucker 1967, p. 35). Hence, the first step to be an effective manager is to learn to manage oneself, and one of the most crucial issues in this is to learn to manage time usage.

7.2 TIME IN MANAGERIAL ENVIRONMENT

Experienced time is not the same for everybody. Time can be divided into two categories: subjective time and objective time (Harung 1998). Objective time is the same for everybody and can be understood as chronological time, business time, where the speed of change is the same for everybody. Subjective time is heavily relativistic, and its speed is dependent on many different factors that affect a person's experience of time and are biasing experience from objective time. These factors are, for example, a person's way of utilizing and sequencing time, feeling (Harung 1998), a person's cultural background (Lewis 2010), situation, time pressure (Kobbeltvedt et al. 2005), lack of sleep (Barnes et al. 2011; Kobbeltvedt et al. 2005), personal traits (Berglas 2004), and planning personality (Buehler and Griffin 2003).

A good example for perceiving the difference between chronological time and experienced time can be found from the ancient Greeks. According to Czarniawska (2004), the Greeks divided time according to gods named Chronos (the god of time) and Kairos (the god of proper time). The difference between these two gods was that when Chronos measured time in mechanical intervals, Kairos "jumped and slowed down, omitted long periods and remained in others." In fact, the Greek word *Kairós* means time, place, and circumstances of a subject. The "Kairos" time is something that everybody has experienced. It is the proper time that people are living and feeling. Everybody has experienced, at some point of their life, a feeling of timelessness (Mainemelis 2001), that is, "time flies." This, in extreme cases, can cause a phenomenon called flow, where a person is completely focused and motivated (Csikszentmihalyi 2000). On the other hand, everybody has experienced feelings when time has nearly stopped when doing something unpleasant or boring.

Cultural background should be taken into account more intentionally nowadays than even a decade ago. There are not many businesses left where managers do not come into a situation where they have to deal with persons from other cultures. Globalized businesses and cross-border activities have guaranteed that almost every corner of the world is covered when working with representatives of other companies. These different cultures are also influencing managers' work. Hence, it is to be considered whether a person is future oriented or present oriented.

Linear time cultures prefer to make one action at time, on time, with a future orientation. Actions are set to a line and executed effectively in a concentrated and punctual way. Multiactive cultures consider that the present reality is more important than future appointments. Accomplishing human transactions is the best use for their time. The core in the perception of multiactive time is to recognize that time is an event and personality related and it could be reformed if needed, irrespective of what the chronological time says. Cyclic time is not linear or personally related, but something that happens over again. In cyclic cultures, there is an unlimited supply of time, even if you personally are not in a position to experience all of it (Lewis 2010, pp. 53–62). Linear time could be compared quite directly to the concept of chronological time. Multiactive time can be compared to the concept of experienced time. Cyclic time is not handled in a C&K application or in this chapter.

People experience the same period of chronological time differently in different kinds of situations. Time is experienced differently depending on the level of satisfaction. A more satisfying situation makes feelings toward time more positive and vice versa (Harung 1998). Turnbull (2004) found that executives appeared to maximize their time utilization in a hectic situation by packing every moment of the day with intensive activities. Turnbull also stated that this kind of hecticness will create an imbalance in the time spent on organizational duties or with family. Leaders use compression as coping with the acceleration in organizational life, that is, dealing with shortening time frames. This means leaving things out while also trying to get to the essence of things. Sabelis (2002) founded that this mindset might speed up the acceleration by "implying that rational reduction of information, emotions and alternatives is necessary to reach organizational and individual goals" (Sabelis 2002, p. 102). This might lead to a situation where attention to activity quality, creativity, open-mindedness, innovations, and empathy is reduced. If we compare this kind of compression of time to Drucker's (1967, p. 31) suggestion, where people "have to feel that we have all the time in the world," it is easily seen that it might affect decisions. Studies show that the balance of personal life and work is the most or second most important attribute of the job and that many would change their job if it would improve the balance between work time and self time (Johnson 2004).

A person's development is also found to be a factor, which makes time a positive thing (Harung 1998). Development activities are connected with job satisfaction and whether a person is future or present oriented. Development activities are always those whose payoffs come in the future. If a person does not receive enough time for discretionary activities and rest and sleep, it may lower their self-control and the possibility of behaving unethically will arise (Barnes et al. 2011). Sleep deprivation can be very harmful in time-pressured activities (Kobbeltvedt et al. 2005).

Personal traits of time abusers fall into four main categories: perfectionist, preemptive, people pleaser, and procrastinator (Berglas 2004). This is strengthened with the idea of Oncken and Wass (1999) regarding delegating skills. People seemed to be characteristically optimistic toward schedule predictions. Optimistic future orientation biases planned schedules more than less optimistic orientation (Buehler and Griffin 2003). Long-term vision reduces bias effects compared to short-term vision (Harung 1998).

Jönsson (2000) proposes that before people can manage their time, they have to work with the concept of time through a four-step metaphor. The first step is to recognize that time can be neither accepted or denied. In the first step, it is seen that time changes and that the changes are irreversible (Sorli and Sorli 2004), and these changes happen whether we like it or not (Reunanen et al. 2012a,b).

The second step is to find systematic ways to become aware of your time and its use. This second step covers the sources of biases, that is, experienced time, cultural differences, and personal traits toward time and the issue of recording it. In the third step, you have your own thoughts and ideas of time with you in the middle, but you are also able to describe these. The third step can be seen when a person can, for example, plan future operations effectively and correctly, by utilizing distributed (outsiders) knowledge and own consciousness predictions and decisions toward time in such a way that it is correct when scrutinized afterward. In the fourth and final step, you can master the concept of time as the possibility to compare and analyze your thinking of time with other methods and thinking processes. The fourth step should be achieved before more demanding processes can be successfully utilized. The fourth step means that people could utilize Kairos-type (proper) time to their own benefit and still not lose control of the Chronos type of synchronization with the environment.

After Jönsson's (2000) fourth step, it is possible to try to handle time in the way that Drucker describes in his three-process time management model (Drucker 2005), where he divides time management into three different processes: recording time, managing time, and consolidating discretionary time into bigger sections. Discretionary time would be the ultimate goal to achieve in time management (Drucker 2005, pp. 225–240). Leaders should also try to clarify who is demanding their time usage. Oncken and Wass (1999) provide a model for boss-imposed time, system-imposed time, and self-imposed time for division model for time usage. Other models are also provided by Bandiera et al. (2011) regarding time spent with outsiders or insiders; Oshagbemi (1995), regarding time spent on deskwork, meetings, giving, or receiving information and by different locations, for example, home, office, and by company time spent; Tengblad (2002), how big portions work is done, that is, for how long leaders are able to concentrate on the task at hand. Oshagbemi (1995), Tengblad (2002), and Bandiera et al. (2011) also studied the length of the working weeks or days of leaders. This should also be recognized when managing time or designing time management models.

This means that when positioning time in a managerial environment for easy to understand metaphor, time could be displayed as in Figure 7.1. This metaphor, called Managerial Windshield, was first introduced by Vanharanta (2008), and further developed by Reunanen (2013).

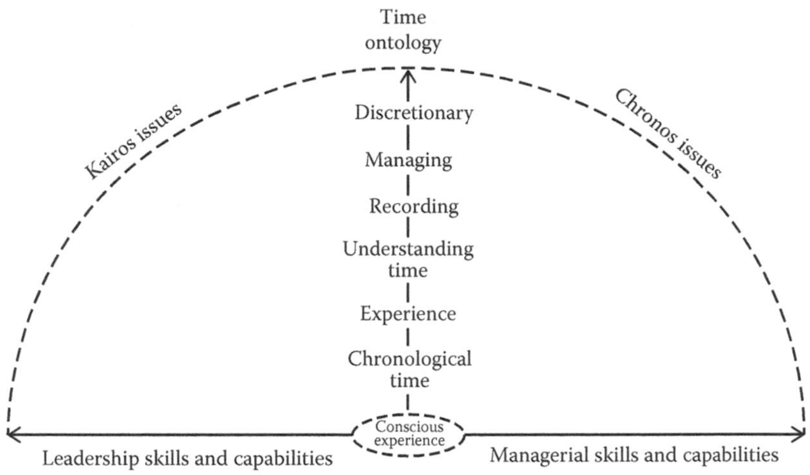

FIGURE 7.1 Time in the Managerial Windshield metaphor. (From Reunanen, T., Leaders' conscious experience towards time, Master of Science thesis, Tampere University of Technology, Pori, Finland, 2013.)

As seen in Figure 7.1, leadership skills and capabilities, that is, issues handled with humans, are related to Kairos-type time issues and managerial skills, and capabilities are related to Chronos-type time, that is, business as a phenomenon is done in easily recorded and divided time, but relations with people involves a much fuzzier and not-so-easy-to-manage time.

7.3 CHRONOS AND KAIROS TOOL

For the evaluation of leaders' conscious experience toward time (Reunanen 2013), a computer application was created in the generic Internet-based Evolute platform, named Chronos and Kairos (Kantola 2005; Kantola et al. 2006). The application was constructed to express time ontology including different (n = 25) features and categories (n = 5). These categories are divided into two main classifications: managing time and experiencing time. The application contains 167 statements to evaluate respondents' current status and their future targets. The idea is to have respondents' feelings transferred into application and translated into numerical data and a visual form.

Here is the hierarchy of categories and features:

- Managing time
  - Understanding time
    - Awareness of time
      - *Time as a resource*: The person is aware of time's special character as a resource. The person sees differences compared to other resources such as money or staff.

    - *Time as an undominated phenomenon*: The person understands that time is a phenomenon beyond human control.
    - *Value for humans*: The person sees time as valuable for human beings.
  - *Recording time*: The person has recorded time in order to benefit from it. The person sees that recording is a way to understand their time usage correctly.
  - *Productivity effectiveness*: The person is aware of which kind of time usage is effective. The person uses their own time in an effective way.
  - *Productivity efficiency*: The person is aware of which kind of time usage is efficient. The person uses time in an efficient way.
  - *Productivity occupancy*: The person is aware of which kind of time usage occupancy is. The person concentrates on one thing at a time.
- Situation
  - Level of haste
    - *Workload*: Level of haste caused by the workload. Person needs to lessen or increase the workload.
    - *Scheduling*: Level of haste caused by wrong scheduling. Persons need to enhance scheduling skills.

- Experienced time
  - Situation
    - Level of haste
      - *Abstract*: Level of haste experienced. Things that are felt but cannot be directly identified.
      - *Concrete*: Concrete things that raise the level of haste. Concrete things that the person can point out.
    - Level of satisfaction
      - *Motivation*: Satisfaction level achieved from the person's motivation
      - *Development*: Satisfaction level gained from the person's development activities
      - *Work–life balance*: Balance between work and free time
    - *Rest*: Possibility to rest enough and recover from work
  - Personal traits
    - *Perfectionist*: Level of perfectionism
    - *People pleaser*: Level of being a people pleaser
    - *Procrastinator*: Level of being a procrastinator
    - *Preemptive*: Level of being preemptive
    - *Optimistic planning personality*: Level of optimistic planning personality
  - Time orientation
    - Discretionary orientation
      - *Free time*: The person's level of orientation toward their own discretionary free time

    - *Working time*: The person's level of orientation toward their own discretionary working time
    - *Thinker time*: The person's level of orientation toward their own thinking time
- *Future orientation*: The person's level of orientation toward the future perspective
- *Present orientation*: The person's level of orientation toward the "here and now" perspective

Statements are modified and developed from studies and models gathered from research field. The sources of knowledge and information are general literature references and also publications from different journals around the world. The sources for journals were mainly from the archives of Elsevier, Emerald, EBSCO host, and *Harvard Business Review*. Backgrounds for statements are naturally references stated in Chapters 1 and 2, but also in various other articles on management and leadership. Leadership research field references prevail because of the vast amount of research on influencing people (Hershey et al. 2001), something that should be felt (Pardey 2007), and changing situations (Mackenzie and Barnes 2007; Vanharanta et al. 1997; Vanharanta and Salminen 2007). Leadership models that considered time were also one of the resources. Models scrutinized were (1) adaptive leadership (Heifetz 1994), (2) change-centered leadership (Skogstad and Einarsen 1998), (3) contingency theory (Fiedler 1964, 1967), (4) LAMPE model (Mackenzie 2006), (5) leader–member exchange theory (LMX) (Northouse 2007; Sherony and Green 2002), (6) multiple linkage model of leadership (Yukl 1981), (7) path goal theory, (8) situational leadership (Hersey and Blanchard 1982; Hersey et al. 2001), (9) team leadership (Mälkki 2009; Northouse 2007; Peck 1990), and (10) transformational leadership (Bass and Avolio 1990).

7.4 CASE

The application's main function is to provide a self-evaluation tool for leaders and a tool for coaching and consulting for managers. The goal is to take the numerical and visual data of personal feelings from the application as guidance to focus on development issues. Even though individual results are the most interesting part of the purpose of Chronos and Kairos, the application could also be used as a research tool for larger groups.

The application was tested in several different cases that could be divided into four larger groups: project managers and professionals in Turku University of Applied Sciences, bachelor's degree students in Turku University of Applied Sciences and master's degree students in Tampere University of Technology, and a heterogenic group of different professionals, students, and other representatives. Many of the tests were performed for certain individuals in order to verify the correct function of the application. Bigger groups were also tested in order to obtain verification for larger data masses handling and data combination. Altogether, 112 individuals were tested between spring 2013 and summer 2014.

7.5 RESULTS

Figure 7.2 shows the results of a heterogenic group study. It shows the combined results of 42 individuals organized into order of creative tension. Creative tension is the term for the difference between current status and future target status. In the histogram in Figure 7.2, the dark gray bar indicates the current status and the light gray bar indicates target status of the feature.

When the light gray bar is longer than the dark gray bar, it means that respondents feel that they want to add this feature in future. When the light bar is shorter than blue bar, it means that respondents feel that they want to lessen this feature. The lengths of bars indicate how strongly respondents feel toward features. Creative tension can be considered to be quite a good meter when issues to be developed are considered. In this particular case, it is seen that this subject group is quite strongly in a hurry and experiencing hectic and fragmented situations. It can be seen, from the figure, that correspondents feel that they need more rest, more free time, and better balance in life in addition to understanding time's nature and time usage in bigger portions (occupancy). At the same time, they feel that they want to lessen their workload, feeling of and concrete hurry and ways how to cope with personal features (procrastination, people pleaser, and perfectionist), which are probably one reason for this situation or at least they are feeling guilty of those.

Figure 7.3 shows the standard deviations of the answers in Figure 7.2. Dark gray and light gray columns show the mean of values with ± standard deviation. The black line shows the mean.

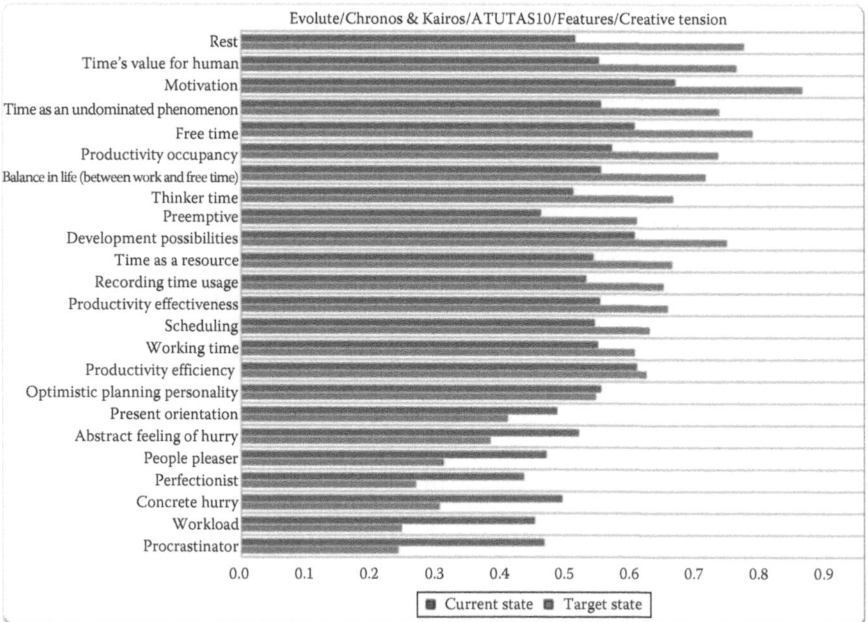

FIGURE 7.2 Heterogenic test group results organized into order of magnitude of creative tension.

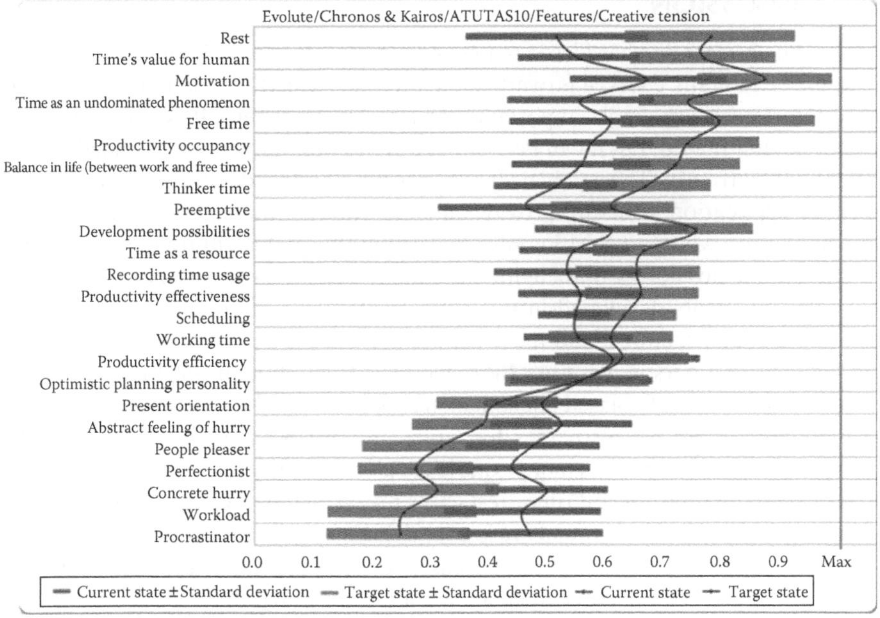

FIGURE 7.3 Heterogenic test group results organized into order of magnitude of creative tension and test group's standard deviation.

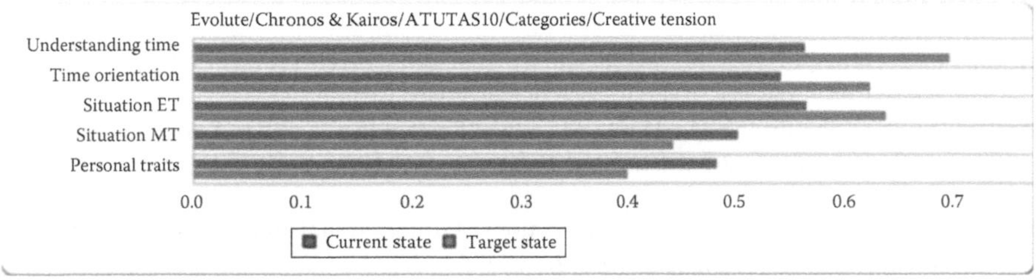

FIGURE 7.4 Heterogenic test group results organized into order of magnitude of creative tension category level.

Since concentration can be observed, this illustration of results gives quite easily possible effective targets for development activities. Figures 7.2 and 7.3 show results on a features level and Figure 7.4 shows results on a category level.

Category level gives the direction of results divided into more of a universal level than features. In Figure 7.4, test group development issues are, on the one hand, mostly understanding the essence and nature of more time and, on the other hand, lessening harmful influence of personal traits on time management. This and other sample studies verify, together with interviews, that the Chronos and Kairos application is considered to be a working self-development tool for decision making regarding leaders' conscious experience toward time and time management.

7.6 DISCUSSION

Time is an abstract and hard issue to understand, but it is something that everybody recognizes. In many cases, it is overlooked too easily and its fundamental basics are forgotten. Time is the most important resource for managers. It is something that cannot be dominated, but it can be exploited well if one's own experienced time is understood correctly. This study deeply bonds time and leadership. The time ontology and application structure provides a clear way for managers to benefit from time and to see where to start in order to develop themselves in time management.

There is the possibility of recognizing leaders' conscious experience toward time by using statements developed in this study. Statements are combined into bigger meaning through features, and these features reveal how leaders feel toward time and how aware a person is toward these feelings. Even though this highly personal and often quite strongly biased phenomenon should always be scrutinized individual by individual, there are also signs for larger guidelines in time management; for example, people who possess more experience also typically have more satisfied and more relaxed feelings toward time. Test cases reveal that it appears that people with a longer professional background are keener to have effective use of their occupancy time than their younger colleagues. Test cases reveal these and many other signs, but these signs should be further studied and strongly scrutinized.

Future research issues could link leaders' time features to features in other research such as organizational commitment, leadership styles, leadership activities, management systems, emotional intelligence, and innovativeness, to mention few.

REFERENCES

Bandiera, O., Guiso, L., Prat, A., and Sadun, R. 2011. What do CEOs do? EUI working paper ECO2011/06. Florence, Italy: European University Institute, p. 23.

Barnes, C.M., Schaubroek, J., Huth, M., and Ghumman, S. 2011. Lack of sleep and unethical conduct. *Organizational Behavior and Human Decision Processes* 115: 169–180.

Bass, B.M. and Avolio, B.J. 1990. The implications of transactional and transformation leadership for individual, team, and organizational development. *Research in Organizational Change and Development* 4: 231–272.

Berglas, S. June 2004. Chronic time abuse. *Harvard Business Review* 82(6): 90–97.

Boroditsky, L. 2000. Metaphoric structuring: Understanding time through spatial metaphors. *Cognition* 75: 1–28.

Buehler, R. and Griffin, D. 2003. Planning, personality, and prediction: The role of future focus in optimistic time predictions. *Organizational Behavior and Human Decision Processes* 92: 80–90.

Csikszentmihalyi, M. 2000. *Beyond Boredom and Anxiety*, p. 231. San Francisco, CA: Jossey-Bass.

Czarniawska, B. 2004. On time, space and actions nets. *Organization* 11: 773–791.

Drucker, P.F. 1967. *The Effective Executive*, p. 178. New York: Harper & Row.

Drucker, P.F. January 2005. Managing oneself. *Harvard Business Review* 83(1): 100–109.

Fiedler, R.E. 1964. A contingency model of leadership effectiveness. In *Advances in Experimental Social Psychology*, pp. 149–190. New York: Academic Press.

Fiedler, R.E. 1967. *A Theory of Leadership Effectiveness*, p. 310. New York: McGraw-Hill.

Harung, H.S. 1998. Reflections. Improved time management through human development: Achieving most with least expenditure of time. *Journal of Managerial Psychology* 13(5/6): 406–428.
Heifetz, R. 1994. *Leadership without Easy Answers*. Cambridge, MA: Press of Harvard University Press.
Hersey, P. and Blanchard, K.H. 1982. *Management of Organizational Behavior: Utilizing Human Resources*, 4th edn. Englewood Cliffs, NJ: Prentice-Hall.
Hersey, P., Blanchard, K.H., and Johnsson, D.E. 2001. *Management of Organizational Behaviour: Leading Human Resources*. Harlow, U.K.: Prentice-Hall Inc.
Johnson, J. 2004. Flexible working: Changing the manager's role. *Management Decision* 24(6): 721–737.
Jönsson, B. 2000. *10 Ajatusta ajasta*. Hämeenlinna, Finland: Karisto Oy.
Kantola, J. 2005. Ingenious management. Doctoral thesis, Pori, Finland: Tampere University of Technology, Publication 568.
Kantola, J., Vanharanta, H., and Karwowski, W. 2006. The Evolute system: A co-evolutionary human resource development methodology. *International Encyclopedia of Human Factors and Ergonomics* 3: 2902–2908.
Kobbeltvedt, T., Brun, W., and Laberg, J.C. 2005. Cognitive processes in planning and judgements under sleep deprivation and time pressure. *Organizational Behavior and Human Decision Processes* 98: 1–14.
Lewis, R. 2010. *When Cultures Collide*, 3rd edn. Helsinki, Finland: WS Bookwell.
Mackenzie, K.D. 2006. The LAMPE theory of organizational leadership. In Yammarino, F.J. and Dansereau, F. (eds.), *Research in Multi-Level Issues: Multi-Level Issues in Social Systems*, vol. 5, pp. 345–428. Oxford, U.K.: Elsevier.
Mackenzie, K.D. and Barnes, F.B. 2007. The unstated consensus of leadership approaches. *International Journal of Organizational Analysis* 15(2): 92–118.
Mainemelis, C. 2001. When the muse takes it all: A model for the experience timelessness in organizations. *Academy of Organizational Review* 26(4): 548–565.
Mälkki, J. 2010. Herrat, jätkät ja sotataito—mutta mikä sai armeijan taistelemaan? (Officers, grunts and military skills—but what made army to fight?) In Siren, T. (ed.) *Maanpuolustuskorkeakoulun Johtamisen ja sotilaspedagogiikan laitoksen Julkaisusarja*, Publication series 2. Helsinki, Finland: National Defence University of Finland, Department of Leadership and Military Pedagogics.
Northouse, P.G. 2007. *Leadership: Theory and Practice*. Thousand Oaks, CA: Sage.
Oncken, W. Jr. and Wass D. L. 2007. *Management Time: Who's Got the Monkey*? Boston, MA: Harvard Business Review. November–December 1999. Reprinted 2007.
Oshagbemi, T. 1995. Management development and managers' use of their time. *Journal of Management Development* 14(8): 19–34.
Pardey, D. 2007. *Introducing Leadership*. Oxford, U.K.: Elsevier Ltd.
Peck, T. 1990. Leadership—A doctrine lost and found. Fort Leavenworth, KS: School of Advanced Military Studies United States Army Command and General Staff College.
Reunanen, T. 2013. Leaders' conscious experience towards time. Master of Science thesis, Pori, Finland: Tampere University of Technology.
Reunanen, T., Valtanen, J., and Windahl, R. 2012a. Evolutionary approach to modern creative engineering studies at Turku University of Applied Sciences. In *Proceedings of the International Conference on Engineering Education*, Turku, Finland, July 30–August 3, 2012.
Reunanen, T., Valtanen, J., and Windahl, R. 2012b. Evolutionary approach to product development projects. In *Proceedings of the METNET Seminar 2012 in Izmir: Metnet Annual Seminar in Izmir*, Izmir, Turkey, October 10–11, 2012.

Sabelis, I. 2002. Hidden causes for unknown losses: Time compression management. In Whipp, R., Adam, B., and Sabelis, I. (eds.), *Making Time*. Oxford, U.K.: Oxford University Press.

Sherony, K.M. and Green, S.G. 2002. Relationships between co-workers, leader-member exchange, and work attitudes. *Journal of Applied Psychology* 87: 542–558.

Skogstad, A. and Einarsen, S. 1998. The importance of a change-centered leadership style in four organizational cultures. *Scandinavian Journal of Management* 15: 289–306.

Sorli, A. 2002. Time as a stream of change. *Journal of Theoretics* 4–6.

Sorli, A. and Sorli, K. 2004. Does time really exist as a fourth dimension of space? *Journal of Theoretics* 6–3.

Tengblad, S. 2002. Time and space in managerial work. *Scandinavian Journal of Management* 18: 543–565.

Turnbull, S. 2004. Perceptions and experience of time-space compression and acceleration. *Journal of Managerial Psychology* 19(8): 809–824.

Vanharanta, H. 2008. The management windshield: An effective metaphor for management and leadership. In Karwowski, W. and Salvendy, G. (eds.), *AHFE International Conference*, Las Vegas, NV, July 14–17, 2008.

Vanharanta, H., Pihlanto, P., and Chang, A.M. 1997. Decision support for strategic management in a hyper-knowledge environment and the holistic concept of man. In *Proceedings 30th Annual International Conference on Systems Sciences*, Maui, Hawaii, pp. 243–258.

Vanharanta, H. and Salminen, T. 2007. Holistic interaction between the computer and the active human being. *Human-Computer Interaction. Interaction Design and Usability Lecture Notes in Computer Science* 4550: 252–261.

Yukl, G.A. 1981. *Leadership in Organizations*. Englewood Cliffs, NJ: Prentice-Hall.

8 Organizational Resource Innovation

Innovation is creating something new or better than before, that is, a process, product, service, technology, system, or any combination of these. Innovations are typically classified by type or category (Subramanian and Nilakanta 1996). In this chapter, we focus on innovation regarding organizational resources. By following the Evolute approach systematically, an organization's knowledge base is expanded and can be used to innovate regarding organizational resources. Nonaka and Takeuchi (1995) describe how interplay between tacit and explicit knowledge contributes to new knowledge creation. The assumption here is that by systematically expanding the knowledge base, the chance to innovate will eventually increase. So-called knowledge increments expand the knowledge base. The Evolute approach makes innovation possible on different knowledge levels. Innovation can be based on different aspects of ontology, ontology development processes, and ontology-based resource management. The scope of innovation varies from a detail to a large system-level innovation. The knowledge increments expand the knowledge base for potential innovation. The following sections describe how knowledge increments contribute to organizational knowledge creation and potential innovation.

Previous sections in this book have illustrated how a knowledge base can be built and developed according to the Evolute approach. Each proposition describes a step that adds something new to the knowledge base, and each step is required to build collective understanding of our world and its state. Figure 8.1 shows how the expansion of the knowledge base is actually done according to the Evolute approach.

All levels in Figure 8.1 are explicit knowledge levels (compare with the definition of instance). However, all levels above ontology require externalization from individuals' tacit knowledge to explicit knowledge. This is one of four knowledge conversions in Nonaka's knowledge creation/socialization, externalization, combination, internalization (SECI) process (Nonaka and Takeuchi 1995), and therefore, the Evolute approach supports the SECI process in an organization. The SECI process itself aims at innovations through new knowledge creation. Table 8.1 describes the scope of knowledge increment and innovation on each level. Incrementally growing the knowledge base allows us to better understand and innovate using the new incremental knowledge pieces and knowledge combinations they create with existing knowledge pieces on the same level and across the levels, that is, on both interknowledge and intraknowledge levels. In this book, all these levels describe organizational resources. The organizational resource innovation refers to understanding the resources, their content and structure, as well as their behavior, better than before, which may thus lead to the creation of new resources and the better arrangement of existing ones. Through organizational resource innovation, we can presume to have better resources than earlier that will turn into better performance by the organization. This causality is difficult to prove, though.

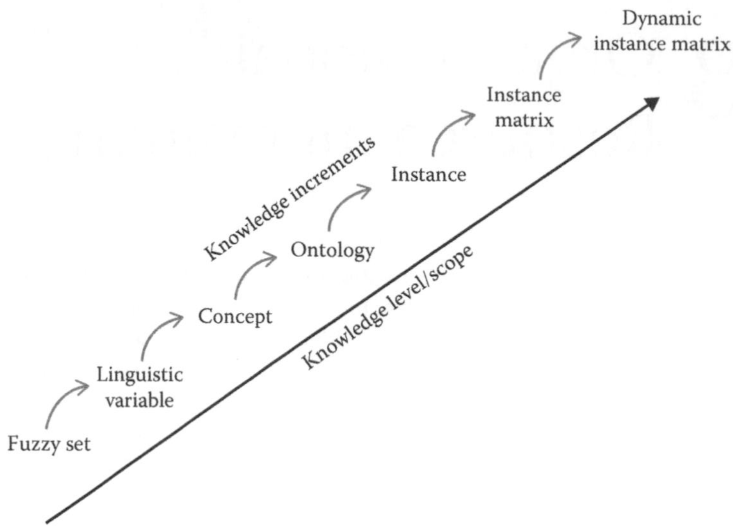

FIGURE 8.1 A knowledge base is expanded on several levels (c.f. Kantola and Vanharanta 2012) according to the Evolute approach.

TABLE 8.1
Knowledge Base Expansion Increases Understanding

Knowledge Scope	What New Aspect Is Added to the Knowledge Base?	What Is Understood Better Than Earlier?
Fuzzy set (membership function)	Value range for the indicator	The behavior of the indicator
Linguistic variable (indicator)	Easily understandable and perceivable viewpoint of the concept	The concept-indicator causality
Concept	Relevant element in the domain/part in the domain system	Organizational resource domain
Ontology	Explicitly specified domain	Organizational resource management portfolio
Instance	The perception of the organizational resource in specific context	Current and future states of the organizational resource
Instance matrix	The collection of instances at one time	Current and future states of organizational resource management portfolio and knowledge asymmetry (Kantola et al. 2012)
Dynamic instance matrix	The collection of instance matrices over a longer period of time	Current and future states of organizational resource management portfolio longitudinally arranged and dynamic knowledge asymmetry (Kantola et al. 2012)

Based on Table 8.1, it can be seen that the added knowledge items increase in a very real way the understanding of an organization's resources which, in turn, will lead to plans and action to develop organizational resources creatively and continuously. Furthermore, this should lead to improved performance of an organization. The chances to innovate are increased by the following:

1. Seeing new knowledge combinations on the intraknowledge and interknowledge levels
2. Bottom-up incremental layered examination and use of knowledge to innovate regarding the ontology in focus
3. Understanding changes from bottom to top and from top to bottom
4. Seeing and understanding knowledge asymmetry
5. Seeing and understanding dynamic knowledge asymmetry—dynamic collective wisdom and history
6. Improving knowledge representation and ontological modeling

Figure 8.2 illustrates dynamic knowledge asymmetry in an organization on an instance matrix level. There are three management cycles 1–2–3 that together

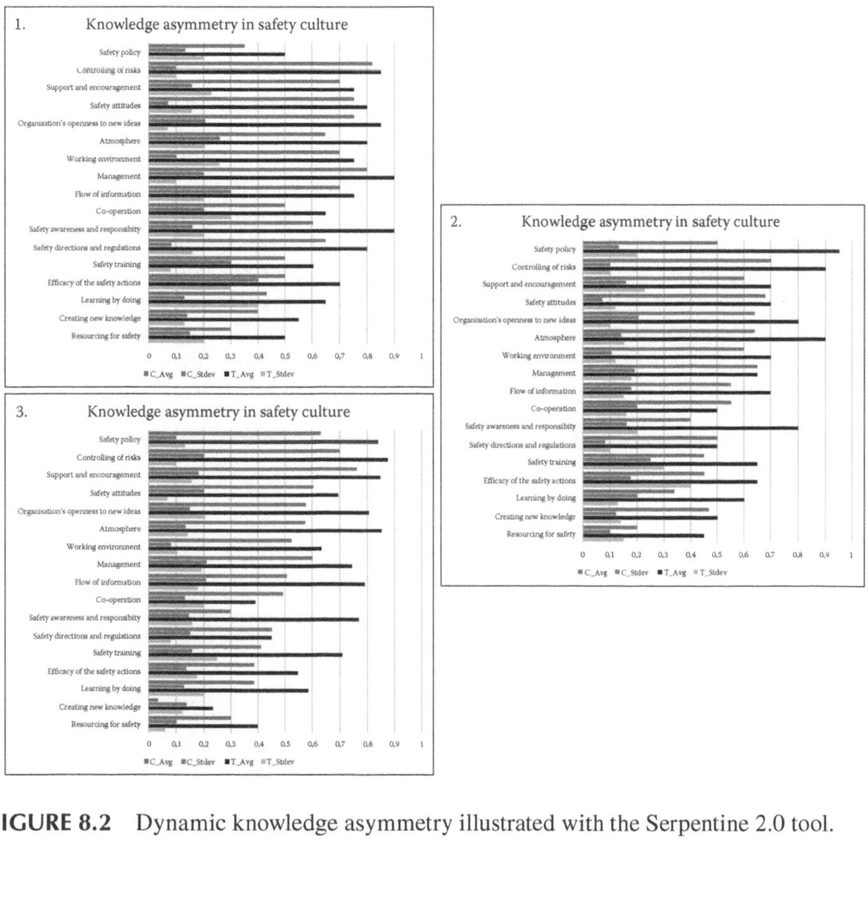

FIGURE 8.2 Dynamic knowledge asymmetry illustrated with the Serpentine 2.0 tool.

cover 18 (1/2 year/cycle) or 36 (1 year/cycle) months in an organization. The collective perception of the current state and future needs clearly varies on each management cycle.

Knowledge bases can be expanded in practice on different levels starting from linguistic variables and culminating in collective perceived dynamics of knowledge domains. Based on the extensive experience of the Evolute approach, it is known that the described steps really work in practice, but it is not yet known whether they actually lead to increased performance and profit. This is difficult to prove. The assumption is that the proposed approach leads to better chances to innovate regarding organizational resources as compared to static knowledge bases. However, these steps and the increased chances to innovate do not automatically lead to increased organizational resource performance. But organizational resource innovation becomes more likely than before. The mechanisms behind potentially new organizational resource innovations are diverse. The proof that these kinds of knowledge increments really lead to increased innovation and performance is needed in the future. If this is the case, then these kinds of new potential benefits will enable innovation processes and encourage innovators in organizations toward better results. In addition, new types of support systems for organizational resource innovation processes, innovation management, and innovators can be set up. In the next chapter, some management methods that are typically used in organizations are discussed.

REFERENCES

Kantola, J. and Vanharanta, H. 2012. Ontologies enable innovation. In *Proceedings of the Seventh European Conference on Innovation and Entrepreneurship (ECIE 2012)*, Santarem, Portugal, September 20–21, 2012.

Kantola, J., Vanharanta, H., Paajanen, P., and Piirto, P. 2012. Showing asymmetries in knowledge creation and learning through proactive vision. *Theoretical Issues in Ergonomics Science* 13(5): 570–585.

Nonaka, I. and Takeuchi, H. 1995. *The Knowledge-Creating Company: How Japanese Companies Create the Dynamics of Innovation*. New York: Oxford University Press.

Subramanian, A. and Nilakanta, S. December 1996. Organizational innovativeness: Exploring the relationship between organizational determinants of innovation, types of innovations, and measures of organizational performance. *Omega* 24(6): 631–647.

9 Other Approaches and Methods

There are many approaches and methods that are directly or indirectly used for the management of resources in organizations. In this chapter, some of the most commonly used approaches and methods are briefly outlined and matched with the Evolute approach. The list of approaches and methods in this chapter is not comprehensive.

Resource-based view (*RBV*) refers to the viewpoint that the competitive advantage of an organization is based on the intangible and tangible resources it has or can utilize (Prahalad and Hamel 1990; Wernerfelt 1984). The resources should have VRIN (valuable, rare, in-imitable, non-substitutable) (Barney 1991) or VRIO (valuable, rare, costly to imitate, organized to capture value) (Rothaermel 2013) attributes to become such resources as can provide competitive advantage to the company. RBV does not provide practical tools to develop and manage the resources so that they have VRIO attributes in the future.

Resource planning in general refers to acquiring, allocating, and releasing resources that are needed in an organization to complete assigned and arising tasks. Management activity related to resourcing involves analyzing situations and scenarios and making those decisions that ensure that the right kind of resources is available at the right time and to the required level. Resource planning systems apply several different kinds of models and methods. Resource planning is guided by the goals of the organization, based on what resources are needed, how much and when, in order to create the target value.

Business process modeling describes processes in the company (Rolstadås 1995). It shows how materials and information flow through tasks and resources in an organization in order to produce products and services. Process models are good at visualizing how work is done now and how it should be done in future in the organization. Top-level process models describe the main processes in an organization, while more detailed subprocess models support the main processes. Required activities, their connections, and necessary resources are described in a visual manner. Process models focus on the flow aspect and do not cover the conceptual side of resources. Management related to business process models aims to ensure that processes are modeled in the proper way and that material and information flows through the processes as planned, that is, *business process management* (*BPM*): see, for example, Jeston and Nelis (2013). *Process simulation*, in turn, entails model-based computing and a what-if type of analysis with regard to how well processes work, and how the processes can be set up.

Common sense management is about making decisions based on one's observations, feelings, and intuitions in a specific situation. This is a risky activity since the situational knowledge that a manager has in the specific situation is typically not as complete as it should be. Common sense fails managers in decision making (cf. Watts 2011). Common sense management can be applied in simple everyday management, but for decision making regarding the resources of an organization, it is clearly not an adequate approach.

Human resource management (*HRM*) has a focus on managing and developing the human resources of an organization. It has two primary views: strategic and operational. The strategic view works together with general top-level management, while the operational view focuses more on everyday HR action: see, for example, Armstrong and Taylor (2013). HRM focuses mainly on one organizational resource type. Following the same logic, *operations management*, *technology management*, *customer relationship management* (*CRM*), etc., likewise focus on certain types of organizational resources. Under these different management themes, the Evolute approach can be very well applied.

Explicit knowledge mining is an activity where internal and external explicit knowledge (Nonaka 1995) sources are searched and analyzed with algorithms to find something hidden that management can use to make decisions: trends and patterns, for example. Explicit knowledge sources cover many different kinds of encoded material, such as numbers, text, pictures, videos, and sounds. But, of course, tacit knowledge cannot be mined from people's minds with algorithms.

Enterprise resource planning (*ERP*) *systems* are software solutions that integrate several business and operation functions of an organization. ERP systems track business resources and business commitment. ERP facilitates information flow between all business functions, and manages connections to outside stakeholders (Bidgoli 2004).

Balanced scorecard (*BSC*) is a framework for performance measurement methods and strategic management that integrates financial, customer, learning, and growth data as well as business process figures with the vision and strategy of an organization (Ahn 2001; Kaplan and Norton 1996). For each set of figures, objectives, measures, targets, and initiatives are defined. A lot of tools related to the concept of BSC are in use in organizations.

Key performance indicators (*KPIs*) are typically numerical measures that are followed in an organization to see how well the organization has performed in terms of business, finance, operations, etc. (see, e.g., Marr 2012; Parmenter 2010). KPIs are compared with set target values to see which aspects the organization has performed well in and which aspects need attention. KPIs can be single variable or multivariable measures—therefore, they are simplistic measures of complex resource systems. These kinds of numerical approaches typically lead to the management of complex resources by scientifically precise numbers and digits on Excel spreadsheets! Clearly, KPIs do not provide an adequate approach to managing organizational resources in real life. Too much stays hidden. Performance is the result of concepts in a domain. The cause is much more attractive than the effect from the management point of view.

These different approaches can be matched with the Evolute approach according to the framework (model, involve, compute, and manage). Let us remember that in the framework in Figure 1.5, each proposition is based on the previous one. It is not enough to make one layer of the framework work—we need to make all the layers of the framework work in the organization. We can see that each approach and method mentioned briefly in this chapter is clearly fundamentally different to the layered organizational resource management framework that is presented in this book. It can be said that currently only the Evolute approach allows all four organizational resource management propositions to become true, to the best of the author's knowledge. There is personal bias by the author involved, of course. However, the four propositions can be thought of as universally true. If so, then the organizational resource management framework presented in this book can guide all kinds of organizations toward more meaning management than in the past.

The Evolute approach is one solution to the framework. It is important to note that the Evolute approach does not intend to replace any of the approaches and methods mentioned in this chapter. Instead, it will provide deeper meaning for the organization in supporting these other approaches and methods by enabling the integration of the most valuable asset organizations have—the experience and understanding of their people. The Evolute approach can be easily integrated with any of these existing approaches and methods.

REFERENCES

Ahn, H. 2001. Applying the balanced scorecard concept: An experience report. *Long Range Planning* 34(4): 441–461.

Armstrong, M. and Taylor, S. April 2014. *Armstrong's Handbook of Human Resource Management Practice*, 13th edn., p. 3. London, U.K.: Kogan.

Barney, J.B. 1991. Firm resources and sustained competitive advantage. *Journal of Management* 17: 99–120.

Bidgoli, H. 2004. *The Internet Encyclopedia*, vol. 1. New York: John Wiley & Sons Inc.

Jeston, J. and Nelis, J. 2013. *Business Process Management—Practical Guidelines to Successful Implementation*, 3rd edn. London and New York: Routledge.

Kaplan, R.S. and Norton, D.P. 1996. *The Balanced Scorecard: Translating Strategy into Action*. Boston, MA: Harvard Business School Press.

Marr, B. 2012. *Key Performance Indicators (KPI): The 75 Measures Every Manager Needs to Know*, 1st edn., Series: Financial Times Series. Harlow, U.K.: FT Press.

Nonaka, I. and Takeuchi, H. 1995. *The Knowledge-Creating Company: How Japanese Companies Create the Dynamics of Innovation*. New York: Oxford University Press.

Parmenter, D. 2010. *Key Performance Indicators (KPI): Developing, Implementing, and Using Winning KPI*, 2nd edn. Hoboken, NJ: Wiley.

Prahalad, C.K. and Hamel G. May–June 1990. The core competence of the corporation. *Harvard Business Review* 68(3): 79–91.

Rolstadås, A. (ed.) 1995. Business process modeling and reengineering. In *Performance Management: A Business Process Benchmarking Approach*, pp. 148–150. Springer-Science+Business Media. B.V.

Rothaermel, F.T. 2012. *Strategic Management: Concepts and Cases*. Boston, MA: McGraw-Hill/Irwin.

Watts, D. March 29, 2011. *Everything Is Obvious: *Once You Know the Answer: How Common Sense Fails Us*. New York: Crown Business.

Wernerfelt, B. April–June 1984. A resource-based view of the firm. *Strategic Management Journal* 5(2): 171–180.

10 Discussion and Future Work

In this book, an approach has been put forward for how to model complex and abstract organizational resources and how to develop and manage these resources for the benefit of the organization and its employees. There is a real need for concrete and holistic development and management work with difficult and abstract concepts in companies and organizations. For that purpose, relevant concepts must be clarified so that they can be understood and used. Though we cannot directly measure abstract resources, we can indirectly ask the people who work with them how they perceive their state and development needs of these concepts through indicators. This is a shared dialogue to manage and develop common resources. Involving people also has a well-known motivational effect. The collective hands-on understanding of organizational resources in changing work situations is very valuable for management decision-making.

This book demonstrated a practical ontology–based resource management framework and approach that helps managers to lead their people in different work roles and manage objects in a targeted way. The approach is generic and can be applied to any domain. The approach is also based on extensive research. There are no limitations regarding the type or content of resources (or objects) that can be managed according to the approach shown in this work. Neither is the approach limited to business management. It is an equally suitable approach for nonprofit management and development in the education, environment, social, and municipal sectors. It is an applicable approach outside of management, too, for instance, in the clarification of mutual issues between stakeholders in collaboration, politics, humanitarian work, etc.

The "quality" of ontologies defines limits. How well do the ontologies describe the real-world objects? A lot of resources are not yet very well specified, and in the future, the more reliable and valid the ontology of the object can be, the bigger the impact the approach will have. In the following sections, some ideas based on this work are presented.

10.1 NEW WORK ROLES

According to the evolute approach, all organizational resources can be developed and managed in a uniform way by following the management cycles shown in Figures 3.31 and 3.32. This suggests that there can be qualified (and licensed) persons to facilitate these cycles. Imagine a certified manager who manages an investment portfolio, the competency training of personnel, and a laundry service simultaneously while following the same method. Such new work roles in the future could be, for example: meaning manager, instance manager, management cycle facilitator, and ontology R&D officer.

10.2 INSTANCE-BASED MEANING INTERPRETERS

The interpreter in this work may represent a new category of ontology applications: instance-based meaning interpreters. We can think of an interpreter between objects and people, and between different groups of people. From this work, we can take the concept of the interpreter and generalize it to cover new areas. Research in this area would potentially bring great benefits purely, because, as shown in this work, we do not always understand the objects we are dealing with well. Therefore, we cannot really understand other people and other cultures well without an instance-based interpreter. Future research on this subject seems attractive since it is in everybody's interest to understand both objects and each other well in different contexts. With the help of instances, this becomes easier than in the past.

10.3 ONTOLOGY FORUMS

The Evolute approach aims at improving integration and interaction. That is the job of effective managers as well (cf. Ackoff 1986). This can be further helped by forums that promote the use of organizational resource ontologies and other domain ontologies as well. The first *Co-Evolute Conference on Human Factors, Business Management and Society HFBMS2015* was held in Las Vegas in July 2015 in association with *AHFE2015* (*Advanced Human Factor and Ergonomics Conference*). This conference series focuses on resource ontologies, the Evolute approach, and other similar approaches as well. These kinds of forum bring together researchers, developers, and practitioners in the field.

10.4 OPEN-SOURCE ONTOLOGIES

As mentioned earlier, the goal is to improve integration and interaction. This can be helped by the promotion of open source ontologies and their use. Such ontologies are open to everybody and their content and structure is developed transparently over time by organizations, groups, and crowds. In such a situation, it is always known what concepts others are talking about, that is, on which concepts the message we are receiving is based. This could even be a requirement for communication in the future. With common sense, this should be everybody's goal. The following sections present potentially interesting views on open source ontologies.

10.4.1 Cultures and Languages

Open-source ontologies can help us to understand why people behave the way they do. People base their thinking and behavior on their perception of objects. When the content and structure is open, people's behavior becomes more transparent than earlier, and therefore, more understandable and logical to other people. There should be fewer language problems in the future due to the use of open-source ontologies. The translation of ontologies into any language can be automated. People can communicate well if their communication is based on common concepts. This really would help to solve some problems related to globalization since the conceptual base of thought in people's minds is unique and hidden. This is one of the reasons for arguments between people. Arguing without a common and shared conceptual base is a waste of time—it will not lead to an agreement or commitment between parties.

10.4.2 Tacit Knowledge

Losing valuable tacit knowledge when employees leave is a reality everywhere. The challenge is to capture and transfer as much of this knowledge as possible to the next generations. Tacit knowledge can be shared between people, but "recipients" still do not know on what conceptual structures the tacit knowledge in the other person's mind is based. If these structures were open source ontologies, tacit knowledge would come in one package together with the structure and content of a domain. In this way, tacit knowledge would become concept-based tacit knowledge that can be tagged with the label of the concept.

TABLE 10.1
Ideas of the Evolute Approach versus Errors in Efforts to Make Organizational Changes

Errors	Evolute Approach Idea
Allowing too much complacency	Avoids complacency, since it is embedded in management systems. It means that change is always already present and can be managed proactively.
Failing to create a sufficiently powerful guiding coalition	Utilizing instances brings together a powerful guiding coalition where people with a bottom-up view are included in the coalition with top-down guidance and control.
Underestimating the power of vision	Vision (creative tension/proactive vision) is always the driving force behind the change requirements. Utilizing instances means that the power of people's vision is not underestimated or undercommunicated.
Undercommunicating the vision by a factor of 10 (or 100 or even 1000)	Metaknowledge provides possibilities to immediately communicate the vision further.
Permitting obstacles to block new vision	Layered organizations disappear. Fewer obstacles are present during change processes than before.
Failing to create short-term wins	People appreciate knowing that their voice is heard. This is a motivating factor and therefore a (short-term) win. The evolute approach is a continuous process, and therefore, it takes some time to apply it and get results with it.
Declaring victory too soon	Victories are first declared after the change process is completed. All participants show how their positions change and move.
Neglecting to anchor changes firmly in the corporate culture	People's (bottom-up) perception of organizational culture is reflected in the instances. The specific ontologies change and are anchored immediately in the organizational culture.

Source: Kotter, P.K., *Leading Change*, Harvard Business School Press, Boston, MA, 1996.

10.5 CHANGE MANAGEMENT

The Evolute approach might help to ease some of the challenges associated with changes in organizations. Table 10.1 outlines these ideas versus common errors in organizational change efforts according to Kotter (1996).

This work is built on the four propositions that together form the framework for organizational resource management. The Evolute approach is one solution for the framework. The author feels that the Evolute approach is connected to existing theory and also has practical relevance. The theoretical contribution is cross-disciplinary with elements from philosophy, management, knowledge management, motivation, neuroscience, systems, modeling, and computing. In addition to a theoretical contribution, this new framework makes a practical contribution since it is evidently functioning in practice. Currently, it is used in many countries, in many universities, and organizations in many different kinds of projects. More and more material is published about the approach. Translations into several languages have been made and are being made. This shows that the new approach is in increasing practical and academic use. Therefore, it appears that this work has introduced a new, solid approach to help managers in their very complex tasks.

The first step to getting started is to admit the need for help in the work of management!

REFERENCES

Ackoff, R.L. 1986. *Management in Small Doses*. New York: John Wiley & Sons Inc.
Kotter, P.K. 1996. *Leading Change*. Boston, MA: Harvard Business School Press.

Index

P

R

S